PRINTED RESONANT PERIODIC STRUCTURES AND THEIR APPLICATIONS

PRINTED RESONANT PERIODIC STRUCTURES AND THEIR APPLICATIONS

Mahesh Abegaonkar
Lalithendra Kurra
Shiban Kishen Koul

CRC Press
Taylor & Francis Group
Boca Raton London New York

CRC Press is an imprint of the
Taylor & Francis Group, an **informa** business

CRC Press
Taylor & Francis Group
6000 Broken Sound Parkway NW, Suite 300
Boca Raton, FL 33487-2742

Printed on acid-free paper
Version Date: 20161019

International Standard Book Number-13: 978-1-4987-8211-1 (Hardback)

Library of Congress Cataloging-in-Publication Data

Names: Abegaonkar, Mahesh, author. | Kurra, Lalithendra, author. | Koul, Shiban K., author.
Title: Printed resonant periodic structures and their applications / Mahesh Abegaonkar, Lalithendra Kurra, Shiban Kishen Koul.
Description: Boca Raton, FL : CRC Press, Taylor & Francis Group, [2016] | Includes bibliographical references and index.
Identifiers: LCCN 2016023237| ISBN 9781498782111 (hardback ; alk. paper) | ISBN 1498782116 (hardback ; alk. paper) | ISBN 9781498782128 (e-book) | ISBN 1498782124 (e-book)
Subjects: LCSH: Microwave integrated circuits. | Strip transmission lines. | Electromagnetism. | Electric filters, Bandpass. | Frequency selective surfaces. | Printed circuits.
Classification: LCC TK7876 .K678 2016 | DDC 621.381/32--dc23
LC record available at https://lccn.loc.gov/2016023237

Visit the Taylor & Francis Web site at
http://www.taylorandfrancis.com

and the CRC Press Web site at
http://www.crcpress.com

Printed and bound in the United States of America by
Edwards Brothers Malloy on sustainably sourced paper

Contents

Preface

The ever-changing requirements of modern communications have resulted in the need for different antenna designs. Some of these requirements are high-gain, broadband, and reconfigurable in terms of frequency, pattern, polarization, etc. This has presented many challenges for designing the microstrip circuits. Many researchers have contributed to solving these challenges by presenting different antenna and circuit designs.

One of the recent methods of controlling antenna and circuit parameters is the use of periodic structures. This particular technique opens many opportunities for antenna and radio frequency (RF) engineers. Some initial uses of periodic structures for electromagnetic applications include gain enhancement in the antenna, sharp band-pass filters, etc. In these applications, the periodic structure is assumed to be infinite in extent and is printed on a different layer than on the actual antenna or circuit. For most applications, however, these infinite structures are not practical.

This book describes the use of periodic structures that are finite in extent for applications in antennas and microstrip circuits. The unit cells of different periodic structures described in this book are resonant in nature, and as a result, different circuits such as band-stop filters, band-pass filters, diplexers, etc., use them. The resonant behavior also helps in obtaining better characteristics for microstrip circuits.

Understanding the behavior of periodic structures is challenging. Most reported techniques are physics oriented, and it is difficult to analyze these structures. This book presents a simple equivalent circuit-based method to understand the behavior of the presented periodic structures.

The purpose of this book is to describe practical applications of resonant periodic structures that are finite in extent, present ideas on building microstrip circuits using periodic structures that are finite in extent, and analyze the periodic structures in terms of equivalent circuit.

Authors

Shiban Kishen Koul earned a BE in electrical engineering at the Regional Engineering College, Srinagar, Jammu and Kashmir, India, in 1977, an MTech in radar and communication engineering in 1979, and a PhD in microwave engineering in 1983 at the Indian Institute of Technology (IIT), Delhi. He is Dr. R.P. Shenoy Astra Microwave Chair Professor at IIT Delhi. He served as deputy director (strategy and planning) at IIT Delhi from 2012 to 2106. He is currently the chairman of Astra Microwave Products Limited, Hyderabad, a major company involved in the development of RF and microwave systems in India. His research interests include RF MEMS, high frequency wireless communication, microwave engineering, microwave passive and active circuits, device modeling, millimeter wave IC design, and reconfigurable microwave circuits, including antennas. Dr. Koul has successfully completed 34 major sponsored projects, 52 consultancy projects, and 54 technology development projects. He is the author or coauthor of 295 research papers, 7 state-of-the-art books, and 3 book chapters. He holds 10 patents and 6 copyrights. Professor Koul is a Fellow of the Institution of Electrical and Electronics Engineers, USA (IEEE), Fellow of the Indian National Academy of Engineering (INAE), and Fellow of the Institution of Electronics and Telecommunication Engineers (IETE). He is the chief editor of *IETE Journal of Research* and associate editor of *The International Journal of Microwave and Wireless Technologies*, Cambridge University Press. He has delivered more than 230 invited technical talks at various international symposia and workshops. He is currently a serving ADCOM member and a member of IEEE MTT Society's technical committees on Microwave and Millimetre Wave Integrated Circuits (MTT-6) and RF MEMS (MTT-21), member of India initiative team of IEEE MTT-S, Advisor Education Committee, Membership Services Regional Co-coordinator Region-10, vice chair Sight Adhoc Committee MTT-S, and MTT-S Speaker bureau lecturer. He served as a distinguished microwave lecturer of IEEE MTT-S for the period 2012–2014. Currently he is distinguished microwave lecturer-emeritus of IEEE MTT-S.

Dr. Koul is the recipient of a gold medal by the Institution of Electrical and Electronics Engineers Calcutta (1977); S.K. Mitra Research Award (1986) from the IETE for the best research paper; Indian National Science Academy (INSA) Young Scientist Award (1986); International Union of Radio Science (URSI) Young Scientist Award (1987); the top Invention Award (1991) of the National Research Development Council for his contributions to the indigenous development of ferrite phase shifter technology; VASVIK Award (1994) for the development of Ka-band components and phase shifters; Ram Lal Wadhwa Gold Medal (1995) from the Institution of Electronics and Communication Engineers (IETE); Academic Excellence award (1998) from the Indian Government for his pioneering contributions to phase control modules for Rajendra Radar; Shri Om Prakash Bhasin Award (2009) in the field of electronics and information technology; Teaching Excellence Award (2012) from IIT Delhi; award for contributions made to the growth of smart material technology (2012) by the ISSS, Bangalore; Vasvik Award (2012) for the contributions made to the area of information, communication technology (ICT); M.N. Saha Memorial Award (2013) from the IETE for the best application-oriented research paper; and IEEE MTT Society Distinguished Educator Award (2014).

Mahesh Abegaonkar earned a PhD in physics (microwaves) at the University of Pune, Maharashtra, in 2002. He worked as a postdoctoral researcher and an assistant professor at Kyungpook National University Daegu, South Korea, from 2002 to 2005. Currently, he is an associate professor at the Centre for Applied Research in Electronics, Indian Institute of Technology (IIT), Delhi. His research activities include microwave and millimeter wave antennas, electromagnetic band gap (EBG) and defected ground structures (DGS), and reconfigurable microstrip circuits. He has been a recipient of Indian National Academy of Engineering's Young Engineer Award in 2008. He has completed several research projects on antennas and periodic structures. He has two International patents to his credits and has coauthored 80 research papers. He is the current secretary and treasurer of IEEE MTT-S Chapter under the Delhi Section.

Lalithendra Kurra earned a BE in electronics and communication engineering at Arunai Engineering College, Tamil Nadu, in 2001 and an MTech in radar and microwave engineering at the College of Engineering, Andhra University, Andhra Pradesh, in 2004. He worked as an assistant professor from 2004 to 2009, at CVR College of Engineering, Hyderabad, Andhra Pradesh. He joined the Centre for Applied Research in Electronics, Indian Institute of Technology (IIT), Delhi, in 2009 as a PhD candidate and earned his PhD in 2015. He is presently a professor at the CVR College of Engineering, Hyderabad, Andhra Pradesh. His main research interests are EBG (electromagnetic band gap) structures, microwave integrated circuits, and planar antennas.

1

Introduction to Periodic Structures in Electromagnetics

Communication systems have become an incredibly important part of daily life. Because multiple communication systems are now being integrated into a single device, they must be reconfigurable in order to reduce space. Reconfigurable filters—important components in any communication system—are also necessary for different applications to keep the system compact. Electromagnetic bandgap (EBG) structures have attracted researchers due to their unique ability to create a bandgap in the transmission characteristics. By using switching devices, these EBG structures are useful in making reconfigurable filters.

This book discusses the development of compact planar EBG structures and their application in microwave integrated circuits (MICs), such as microstrip filters (bandstop and bandpass) and antennas. The objectives of this book include the following:

1. To develop a compact EBG structure. A planar EBG structure is selected over a mushroom EBG structure as it does not require vias and is easy to fabricate.
2. To characterize the EBG structure using electromagnetic (EM) simulations supported by measurements and propose a method to easily integrate it in microwave circuits.
3. To develop an equivalent circuit model for the proposed EBG unit cell to estimate bandgap position.
4. To employ the proposed EBG structure as a notch filter in ultra-wideband (UWB) filter applications. This bandstop filter is used to develop single and dual band-notched UWB filters and antennas.
5. To study the EBG structure's frequency-selective characteristics and apply them to a patch antenna for gain enhancement.
6. To develop reconfigurable EBG structures by modifying the unit cell or changing the effective inductance/capacitance of the unit cell through switching devices or varactor diodes.

Chapter 1 presents the scope and objective of the book. Section 1.1 defines EBG structures and provides a brief historical background, beginning with its origin. Different types of EBG structures, including miniaturized mushroom and planar EBG structures, and their properties and applications in practical circuits are discussed. Various reported reconfigurable EBG structures are also summarized in this chapter.

Chapter 2 is devoted to the proposed planar EBG structure and its characterization. Section 2.2 explains the geometry of the proposed EBG structure and its evolution from the conventional planar structure. In Section 2.3, the EBG structure is characterized using a dispersion diagram, utilizing the Eigenmode Solver of the CST Microwave Studio software, and the results are compared with the conventional uniplanar compact EBG (UC-EBG) structure. In Section 2.4, the EBG structure is characterized by the surface wave measurement technique proposed by Sievenpiper et al. [11]. In Section 2.5, a one-dimensional array of the proposed EBG structures is coupled to the microstrip transmission line to achieve a bandstop filter (BSF) utilizing the EBG's bandgap property. Finally, in Section 2.6, equivalent circuits of the EBG structure and BSF are proposed, and the modeled results are compared with EM simulated results.

Chapter 3 is devoted to the application of EBG BSFs. Section 3.2 presents the well-known multiple-mode resonator (MMR) UWB filter. In Section 3.3, single and dual band-notched UWB filters are demonstrated by cascading the EBG BSF with the MMR UWB filter. The band-notched UWB antenna is demonstrated in Section 3.4.

Chapter 4 discusses reconfigurable and tunable EBG structures. In Section 4.2, reconfiguration is achieved by connecting an additional structure to the EBG unit cell using PIN diodes. Reconfigurable structure (EBG Unit cell along with the additional structure) is used for switchable band-notched UWB filter application. Section 4.3 discusses tunable BSFs that use varactor diodes. Section 4.4 proposes equivalent circuits of the frequency-switchable unit cell, and the modeled results are verified with EM simulations. At the end of the chapter, Section 4.5 proposes bandwidth reconfigurable EBG BSFs using a simplified unit cell and PIN diode switches.

Chapter 5 examines the development of a bandpass filter (BPF) and diplexer. In Section 5.2, broadside coupling of EBG structures to a microstrip line in a multilayer design is investigated and compared with edge-coupled EBG structures. In Section 5.3, harmonic-suppressed BPFs are developed using broadside-coupled EBG cells, while Section 5.4 discusses a diplexer based on this idea.

Chapter 6 discusses an antenna application using EBG structures. In Section 6.2, the frequency-selective surface (FSS) property of the EBG structure is investigated using CST Microwave Studio to observe the transmission across the EBG structure. Bandpass response is observed to be centered around 10.8 GHz. In Section 6.3, an FSS screen of a 13 × 13 array of the proposed EBG cells is used as a superstrate over a patch antenna operating at 10.8 GHz to enhance antenna gain. This chapter also discusses other applications of periodic structures.

1.1 EBG Structures

EBG structures are periodic structures that prohibit/assist the propagation of specific frequency bands of EM waves for all incident angles and polarization states [1]. These periodic structures can be spatial arrangements of metallic or nonmetallic elements in one, two, or three dimensions. Because the concept of periodic elements originated from the optical domain [2], these elements were called photonic bandgap (PBG) structures (as "photon" refers to light). The dimensions were later scaled to operate at microwave and millimeter-wave frequencies. Because the principle of bandgap applies to all EM waves, the name "electromagnetic bandgap" was proposed. The PBG structure is an optical analogy of a semiconductor in which atoms or molecules of the semiconductor crystal are replaced by a macroscopic media with a periodic dielectric function. As the periodic atomic potential produces energy gaps between electrons, a similarly periodic dielectric function in the PBG material produces gaps [3]. When the periodicity follows the Bragg condition ($p = \lambda_g/2$), the constructive and destructive interference of EM waves results in transmission and reflection bands of wave propagation.

Pure dielectric and metallic–dielectric are two types of EBG structures [4]. The dielectric EBG structures are fabricated by drilling holes in the dielectric medium [5–9] or by arranging square dielectric bars periodically in three dimensions (e.g., a woodpile dielectric structure) [10]. Metallic–dielectric EBG structures are fabricated by etching a periodic metal pattern on an electroplated dielectric substrate. These periodic metal patterns form small resonant blocks, which can be categorized in two ways depending on their geometry: mushroom EBG structures with vias [11,12] or planar EBG structures without vias [13,14]. Metallic–dielectric EBG structures can also be created by etching uniform circles on the ground plane of a microstrip line, as done in References 15–17. In References 18 and 19, nonuniform EBG structures utilizing nonuniform circles on the ground plane are reported to improve the stopband bandwidth and passband ripple. This book concentrates primarily on planar EBG structures.

1.1.1 Mushroom EBG Structure

In 1999, Sievenpiper et al. developed the mushroom EBG structure [11,12]. This structure consists of metal patches on top of a substrate in a two-dimensional lattice separated by a gap; each patch is connected to the ground at the bottom of a substrate by vias.

Figure 1.1a shows a 3 × 3 array of mushroom EBG cells. Origins of the capacitance and inductance in the mushroom EBG cells are shown in the cross-sectional view in Figure 1.1b. Because the patches are separated by a gap, the fringing electric field between the neighboring patches provides the required capacitance. The current flow that runs between the patches through vias provides an inductance in parallel to the capacitance, thus forming a parallel resonant LC circuit. At the resonance frequency, the parallel LC circuit presents high impedance, blocking the currents along the surface at this frequency.

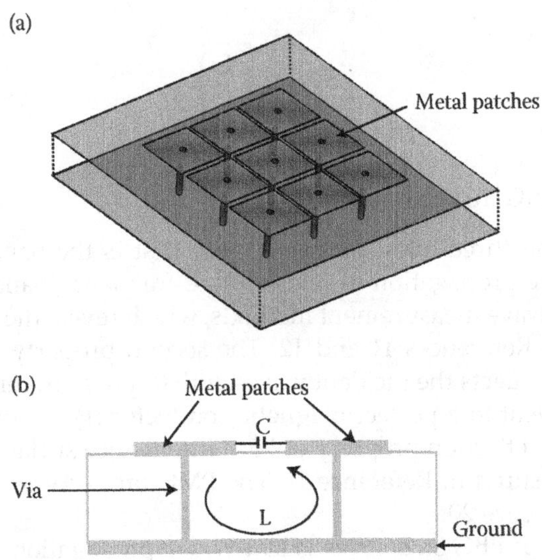

FIGURE 1.1
A mushroom EBG structure: (a) array and (b) cross-sectional view (origin of capacitance and inductance). (D. Sievenpiper et al., Highimpedance electromagnetic surfaces with a forbidden frequency band, *IEEE Trans. Microw. Theory Tech.*, vol. 47, no. 11, pp. 2059–2074, Nov. © 1999 IEEE.)

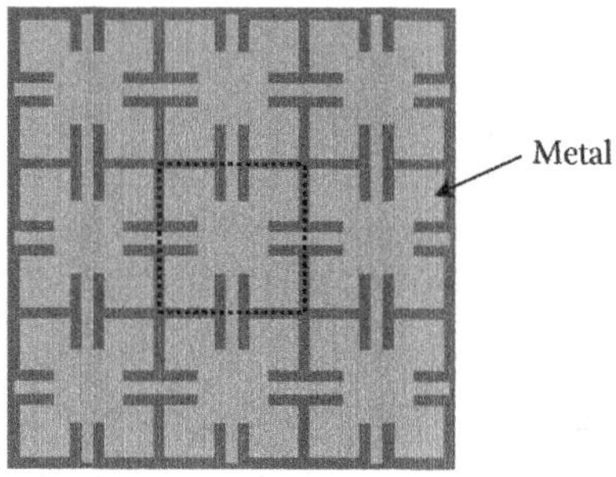

FIGURE 1.2
3 × 3 array of UC-PBG unit cells.

1.1.2 Uniplanar EBG Structure

Yang et al. developed a uniplanar compact PBG (UC-PBG) structure in 1999 [13,14]. A 3 × 3 array of a UC-PBG unit cell is shown in Figure 1.2 (the unit cell is indicated by the dotted line). The unit cell is formed by etching the metal pattern on top of a grounded substrate. Each unit cell is connected to the adjacent unit cells by narrow branches with insets in the metal pad. The current flowing through these narrow branches introduces the inductance, and the gap between the adjacent pads introduces the capacitance. The capacitance and inductance form a parallel resonance circuit. The resonance frequency is inversely proportional to the inductance and capacitance of the unit cell (as given by Equation 1.1), and the bandwidth is given by Equation 1.2. Since planar EBGs do not have vias, they are easy to fabricate.

$$f \propto \frac{1}{LC} \tag{1.1}$$

$$BW \propto \sqrt{\left(\frac{L}{C}\right)}. \tag{1.2}$$

1.1.3 Properties of EBG Structures

EBG structures exhibit three main properties: the first is the bandgap property, which prohibits surface wave propagation in one or more frequency bandgaps. The dispersion diagram and surface wave measurement methods, which reveal the EBG structure's bandgap, are described in References 11 and 12. The second property is the in-phase reflection. An EBG surface reflects the incident waves with 0° phase in certain frequency bands; thus, they are equivalent to a perfect magnetic conductor (PMC) or an AMC for an incident plane wave. The reflection property of high-impedance surfaces (mushroom EBG) is investigated and measured in Reference 11. The PMC property of a UC-PBG structure is demonstrated in Reference 20.

The third property of EBG structures is slow-wave propagation support. A slow-wave factor that is 1.2–2.4 times larger than a conventional transmission line is demonstrated in Reference 21 with a UC-EBG structure etched on the ground plane of a microstrip line. UC-EBG structure is then used in the miniaturization of a BPF in Reference 13 and a patch antenna in Reference 22.

1.1.4 Compact EBG Structures

The conventional mushroom EBG structure [11] and the uniplanar EBG structure [13,14] are popular compact structures that have a wide variety of applications in microwaves. Research has moved toward miniaturization of these EBG structures. Because the bandgap of the EBG structure depends on the effective inductance and capacitance of the unit cell, miniaturization is achieved by increasing the inductance or capacitance of the unit cell.

While the current flow through the via contributes to the inductance in a mushroom EBG structure, etching a spiral on the top plate of a mushroom EBG and connecting to the via provides even more inductance, leading to a compact EBG structure [23,24]. The spiral EBG unit cell is shown in Figure 1.3. Using polar curves and mapping functions, a compact planar EBG is developed in Reference 25, and its schematic diagram is shown in Figure 1.4.

Effective capacitance can be increased by stacking mushroom EBG structures so that two layers of patches overlap each other, thereby providing additional capacitance [12]. The hexagonal array of the overlapped mushroom EBG structures is shown in Figure 1.5.

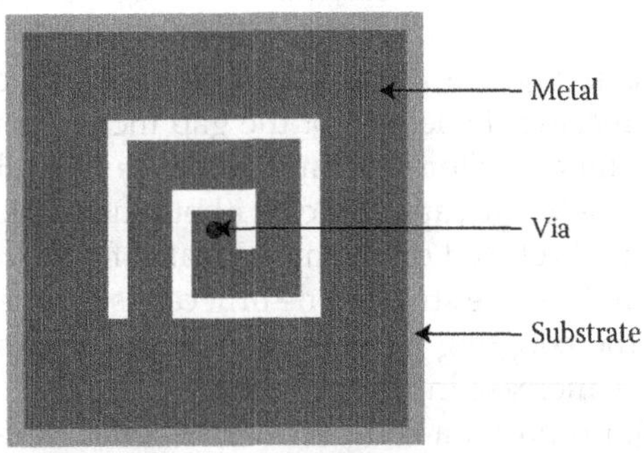

FIGURE 1.3
Spiral mushroom EBG structure. (Adapted from L. Yang, M. Fan, and Z. Feng, A spiral electromagnetic bandgap (EBG) structure and its application in microstrip antenna arrays, in *Asia-Pacific Microwave Conference Proceedings (APMC)*, Suzhou, vol. 3, Dec. 4–7, 2005.)

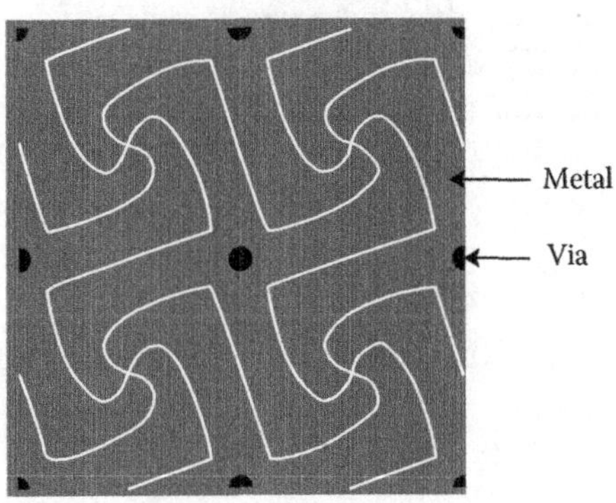

FIGURE 1.4
Polar mushroom EBG structure. (C. B. Mulenga and J. A. Flint, Planar electromagnetic bandgap structures based on polar curves and mapping functions, *IEEE Trans. Antennas Propag.*, vol. 58, no. 3, pp. 790–797, Mar. © 2010 IEEE.)

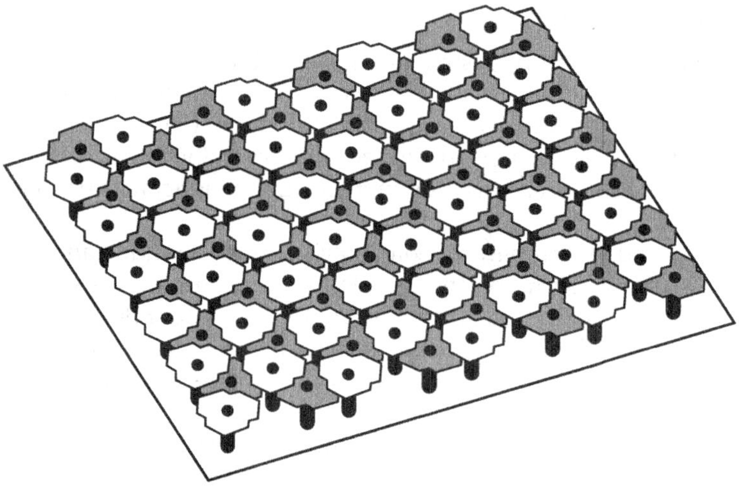

FIGURE 1.5
Stacked mushroom EBG structure. (Adapted from D. Sievenpiper, High-impedance electromagnetic surfaces, PhD dissertation. Dept. Elect. Eng., University of California at Los Angeles, Los Angeles, 1999.)

Because the gap between adjacent metal plates provides the capacitance in the mushroom EBG structure, increasing the length of the gap increases the capacitance. Thus, in Reference 26, an interdigital capacitor is formed with the two adjacent cells (as shown in Figure 1.6) and increases the fringe capacitance without increasing the size of the unit cell.

A fork-like EBG structure is etched on the metal plate of a conventional mushroom EBG structure (shown in Figure 1.7). The strip of one unit cell is inserted into the slot of another cell, thereby increasing the coupling gap between the two cells, which in turn leads to miniaturization due to the increase in capacitance [27].

Rajo-Iglesias et al. [28], proposed a compact mushroom EBG structure, approximately 20% size reduction is achieved as compared to the conventional mushroom EBG by changing the location of the via from center to the edge of the metal patch.

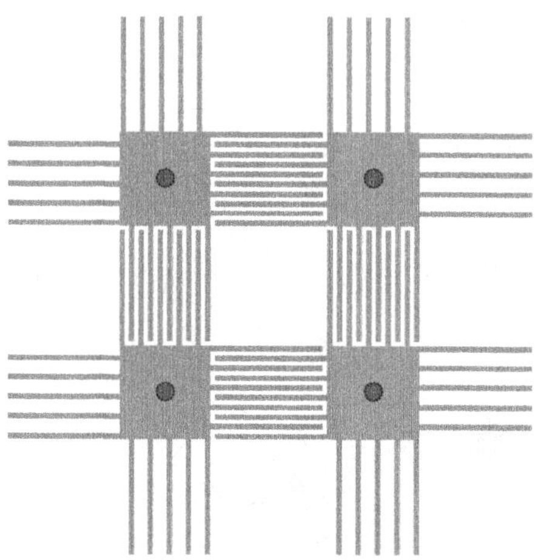

FIGURE 1.6
Mushroom EBG structure with interdigital lines. (Adapted from Y. Fu, N. Yuan, and G. Zhang, Compact high-impedance surfaces incorporated with interdigital structure, *Electron. Lett.*, vol. 40, no. 5, pp. 310–311, Mar. 2004.)

FIGURE 1.7
Fork-like EBG structure. (L. Yang et al., A novel compact electromagnetic-bandgap (EBG) structure and its applications for microwave circuits, *IEEE Trans. Microw. Theory Tech.*, vol. 53, no. 1, pp. 183–190, Jan. © 2005 IEEE.)

By etching a complementary split ring resonator (CSRR) on the patch of the conventional mushroom EBG structure (as shown in Figure 1.8), compactness was achieved in Reference 29. CSRRs significantly increase capacitance and thus decrease resonance frequency without increasing cell size.

Reference 30 presents a compact planar unit cell that is developed by distorting the conventional UC-EBG structure. The distorted unit cell is shown in Figure 1.9. Because there are more conducting lines than in a conventional UC-EBG structure, the additional magnetic flux lines around the conducting lines provide more inductance, which makes the structure more compact.

In Reference 31, the UC-EBG cell is modified as shown in Figure 1.10 to make the structure compact. Because the area around the cell diagonal is filled with triangular metal plates and the meander lines replace the narrow lines of the UC-EBG structure, the effective inductance increases. A thin conductor trace is placed around the unit cell and connected

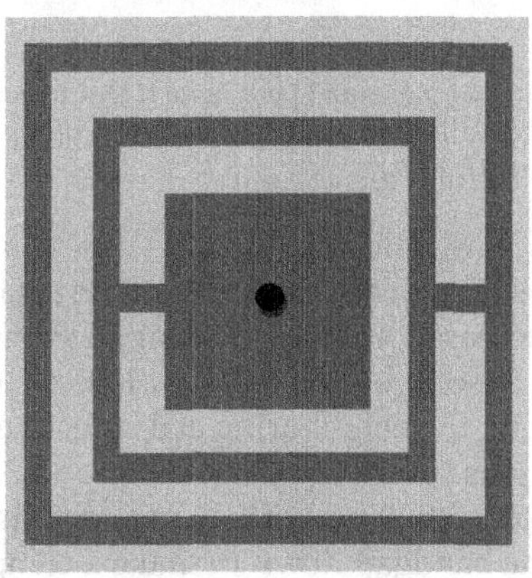

FIGURE 1.8
Compact mushroom EBG structure with CSRR etched on the top metal plate. (L. Peng, C.-L. Ruan, and Z.-Q. Li, A novel compact and polarization-dependent mushroom-type EBG using CSRR for dual/triple-band applications, *IEEE Microw. Wirel. Compon. Lett.*, vol. 20, no. 9, pp. 489–491, Sep. © 2010 IEEE.)

FIGURE 1.9
Compact planar EBG unit cell formed by distorting conventional UC-EBG. (B.-Q. Lin, Q.-R. Zheng, and N.-C. Yuan, A novel planar PBG structure for size reduction, *IEEE Microw. Wirel. Compon. Lett.*, vol. 16, no. 5, pp. 269–271, May © 2006 IEEE.)

FIGURE 1.10
Compact planar EBG cell with triangular plates and peripheral traces. (Adapted from Z. Z. Abidin et al., Design and analysis of UC-EBG on mutual coupling reduction, in *Loughborough Antennas Propagation Conference (LAPC)*, Loughborough, pp. 693–696, Nov. 16–17, 2009.)

to the center, which produces fringe capacitance with the adjacent cell. The effective inductance and capacitance are increased, thereby making the structure compact.

Because the inductance in the conventional planar EBG structure (Figure 1.2) is provided by the narrow lines connecting the neighboring cell, effective inductance can be increased by increasing the length of this line. In References 32 and 33, the effective inductance is increased by replacing the narrow line with a meander line (as shown in Figures 1.11 and 1.12), thereby shifting the bandgap down to make a compact EBG structure.

Even in planar EBG structures like mushroom structures, the capacitance is introduced by a gap between adjacent cells. Increasing the length of this gap using interdigital lines increases the capacitance. A compact EBG structure that incorporates an interdigital capacitor and meander line inductor is presented in Reference 34 and shown in Figure 1.13.

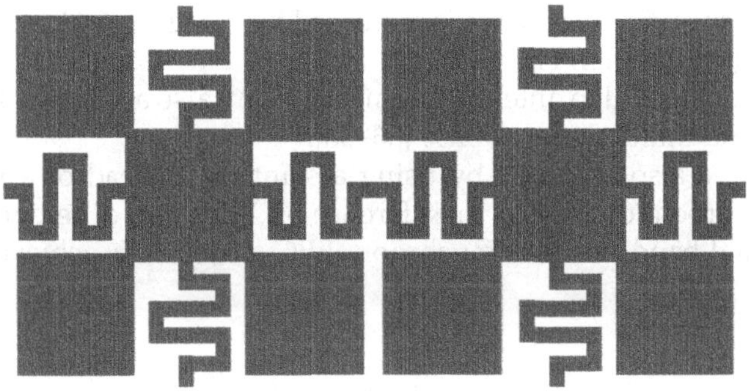

FIGURE 1.11
Schematic of two unit cells of a compact planar EBG structure. (M. F. Abedin, M. Z. Azad, and M. Ali, Wideband smaller unit-cell planar EBG structures and their application, *IEEE Trans. Antennas Propag.*, vol. 56, no. 3, pp. 903–908, Mar. © 2008 IEEE.)

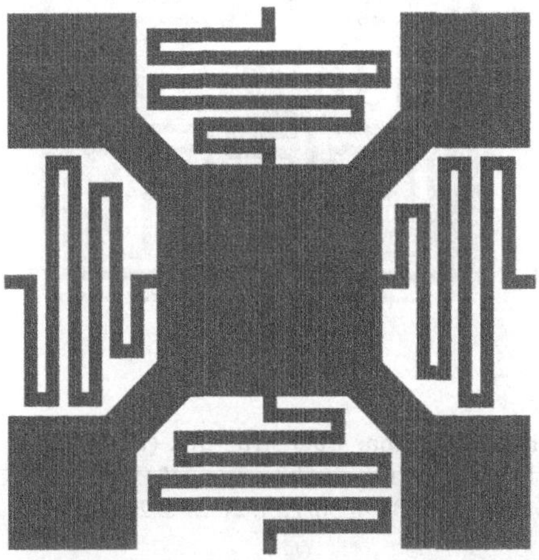

FIGURE 1.12
Compact EBG unit cell. (Adapted from K. Herbertz and S. Lucyszyn, Two-dimensional metamaterials for dual-band filter applications, in *38th European Microwave Conference (EuMC)*, Amsterdam, Oct. 28–30, 2008, pp. 1366–1369.)

FIGURE 1.13
EBG unit cell with meander lines and interdigital capacitors. (Adapted from B.-Q. Lin et al., A novel compact and wide-band uni-planar EBG structure, *Prog. Electromagn. Res. C*, vol. 1, pp. 37–43, 2008.)

However, in Reference 35, capacitance is increased by using a spiral capacitor and a meander line inductor (as shown in Figure 1.14).

Fractals, which are useful in antenna engineering, are also applied to EBG structures for miniaturization and multiband response [36–38].

Miniaturization is also achieved by using a slant via instead of a vertical via in the conventional mushroom dual-band in Reference 39. With two different EBG lattice units, a rabbet spiral dual-band compact mushroom EBG structure is achieved in Reference 40 (shown in Figure 1.15).

FIGURE 1.14
Spiral capacitor and meander inductor planar EBG structure. (W. Wang et al., A novel compact uni-planar electromagnetic bandgap (UC-EBG) structure, in *International Conference on Microwave and Millimeter Wave Technology (ICMMT)*, Nanjing, vol. 4, pp. 1634–1636, Apr. 21–24 © 2008 IEEE.)

FIGURE 1.15
Rabbet spiral dual-band EBG structure. (Y. Yao, X. Wang, and Z. Feng, A novel dual-band compact electromagnetic bandgap (EBG) structure and its application in multi-antennas, in *IEEE Antennas and Propagation Society International Symposium*, Albuquerque, pp. 1943–1946, July 9–14 © 2006 IEEE.)

1.1.5 Application of EBG Structures

Surface waves occur because of the reflections between the substrate ground plane and the air–dielectric boundary at the top of the substrate. These surface waves travel to the edges of the substrate and then diffract, which increases the waste of power in the back lobes. This results in a decrease in the antenna gain and efficiency. However, this problem can be solved by using EBG structures, as they suppress surface waves in the bandgap regions. In Reference 11, the beneficial effects of surface wave suppression are applied to a vertical monopole and a patch antenna to improve their performances. Performance of an aperture-coupled patch antenna on a thick high-dielectric-constant substrate is improved by using a UC-EBG structure to suppress the surface waves [14]. Additionally, in References 41 and 42, suppression of surface waves resulted in gain enhancement and back lobe reduction. Figures 1.16 and 1.17 show patch antennas surrounded by a square lattice of mushroom and planar EBG structures to suppress surface waves.

Another disadvantage of surface waves is that they increase the mutual coupling between antenna elements and circuit components. By using mushroom EBG structures in between antenna elements, mutual coupling is reduced [23,43,44]. Mutual coupling is also reduced by using UC-EBG structures [31,45–47]. In Reference 48, stacked ring patches are used to reduce mutual coupling, whereas in Reference 49 defected ground structures (DGSs) composed of concentric circular rings that provide wide stopband are used to suppress mutual coupling in circular microstrip antenna arrays. Figures 1.18 and 1.19 show mutual reduction in antenna arrays using mushroom and UC-EBG structures, respectively.

A transverse electromagnetic (TEM) waveguide using a uniplanar compact photonic bandgap (UC-PBG) structure is presented in Reference 50. The schematic diagram of the TEM waveguide with UC-PBG is shown in Figure 1.20. A relatively uniform field

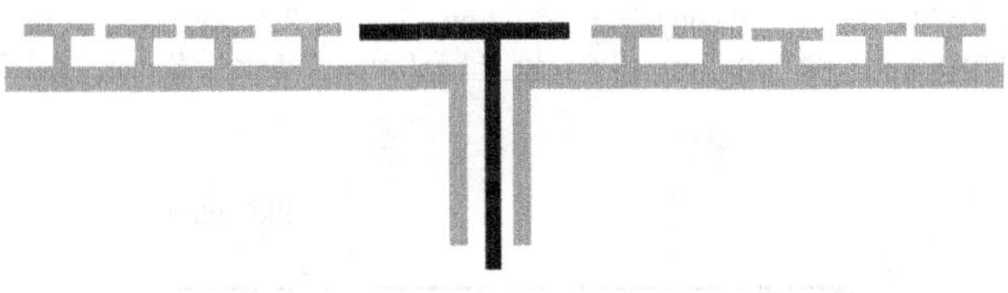

FIGURE 1.16
Cross-sectional view of patch antenna surrounded by mushroom EBG structure. (D. Sievenpiper et al., Highimpedance electromagnetic surfaces with a forbidden frequency band, *IEEE Trans. Microw. Theory Tech.*, vol. 47, no. 11, pp. 2059–2074, Nov. © 1999 IEEE.)

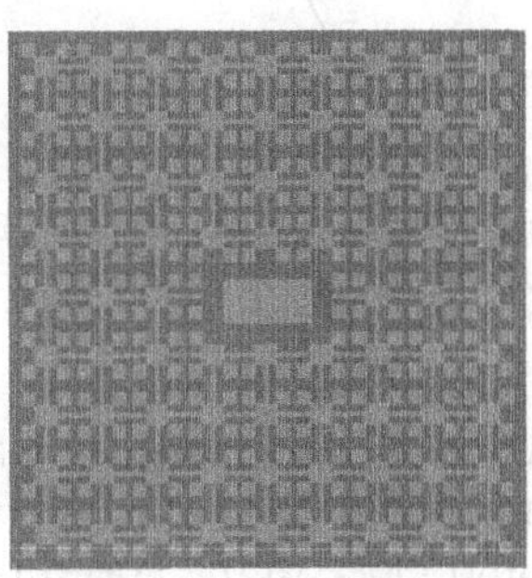

FIGURE 1.17
Top view of patch antenna surrounded by conventional UC-EBG structure. (R. Coccioli et al., Aperture-coupled patch antenna on UC-PBG substrate, *IEEE Trans. Microw. Theory Tech.*, vol. 47, no. 11, pp. 2123–2130, Nov. © 1999 IEEE.)

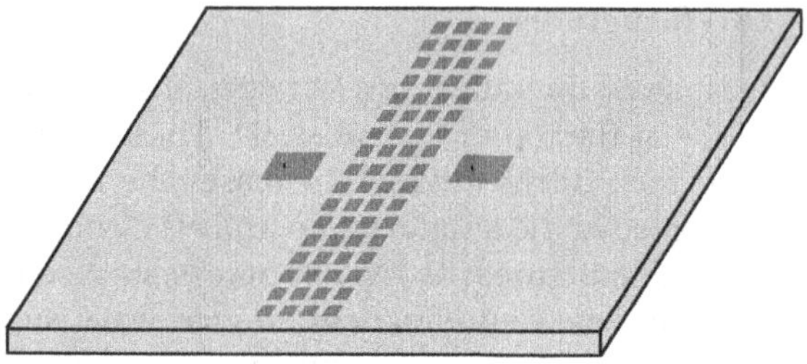

FIGURE 1.18
Array antenna with mushroom EBG for mutual coupling reduction. (F. Yang and Y. Rahmat-Samii, Microstrip antennas integrated with electromagnetic band-gap (EBG) structures: A low mutual coupling design for array applications, *IEEE Trans. Antennas Propag.*, vol. 51, no. 10, pp. 2936–2946, Oct. © 2003 IEEE.)

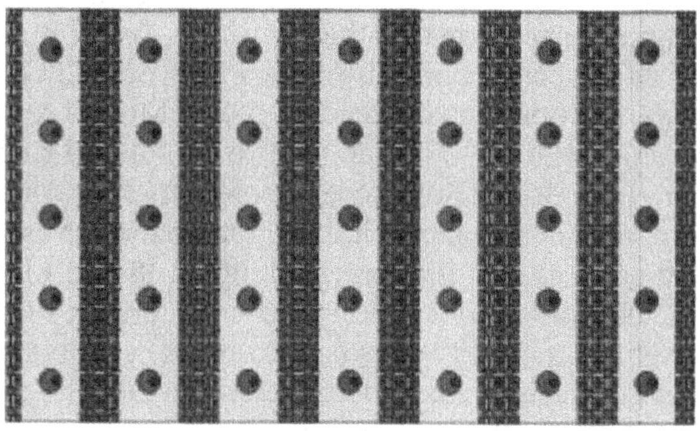

FIGURE 1.19
Array antenna with UC-EBG for mutual coupling reduction. (Z. Iluz, R. Shavit, and R. Bauer, Microstrip antenna phased array with electromagnetic bandgap substrate, *IEEE Trans. Antennas Propag.*, vol. 52, no. 6, pp. 1446–1453, June © 2004 IEEE.)

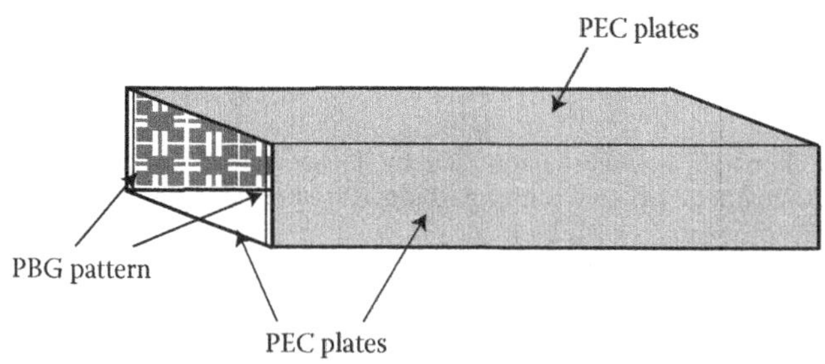

FIGURE 1.20
TEM waveguide with UC-PBG structure. (F.-R. Yang et al., A novel TEM waveguide using uniplanar compact photonic-bandgap (UC-PBG) structure, *IEEE Trans. Microw. Theory Tech.*, vol. 47, no. 11, pp. 2092–2098, Nov. © 1999 IEEE.)

distribution along the cross section of a rectangular waveguide is achieved by placing a PMC using UC-EBG on two narrow walls of the rectangular waveguide. The reported phase velocity is relatively flat and close to the speed of light between 9 and 10.2 GHz, indicating a TEM mode.

A stepped-impedance low-pass filter (LPF) with a UC-PBG structure etched on the ground plane exhibits a very broad stopband, which suppresses the spurious passband

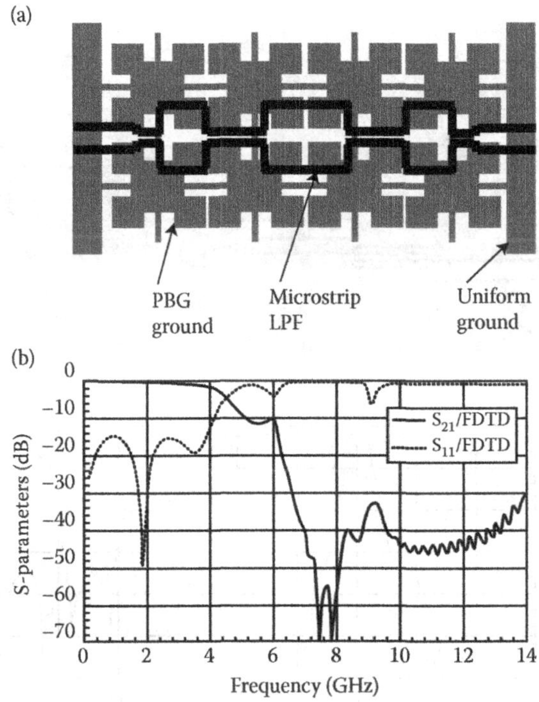

FIGURE 1.21
LPF on UC-PBG ground plane: (a) schematic diagram and (b) results. (F.-R. Yang, Y. Qian, and T. Itoh, A novel uniplanar compact PBG structure for filter and mixer applications, in *IEEE MTT-S International Microwave Symposium Digest*, Anaheim, vol. 3, pp. 919–922, June 13–19 © 1999 IEEE.)

present in LPFs with uniform ground planes [51]. The schematic of an LPF with a UC-PBG ground plane is shown in Figure 1.21a, and its results are shown in Figure 1.21b. An LPF with a UC-PBG ground plane is used in a drain mixer for higher conversion gain and lower local oscillator leakage [51].

A conventional parallel coupled microstrip BPF exhibits spurious harmonics. The UC-PBG ground plane suppresses the spurious harmonics of the BPF as they fall in the wide stopband of the UC-PBG structure. Moreover, the slow-wave effect of the UC-PBG structure makes the BPF filter more compact with a UC-PBG ground plane rather than a conventional ground plane [13]. A BPF with UC-PBG ground planes and its results are shown in Figure 1.22.

The second harmonic in a power amplifier is suppressed by incorporating EBG, thereby improving power amplifier efficiency [52]. In Reference 53, harmonic rejection in the antenna was achieved at its feed level using DGSs.

The conductor-backed coplanar waveguide (CB-CPW) has an additional ground plane at the bottom of the substrate, which forms a parallel plate waveguide with the top ground planes and thus allows a power leakage though the parallel plate mode. By using a UC-PBG ground plane at the bottom of the substrate in the CB-CPW, the leakage can be suppressed in the stopband of the UC-PBG structure.

Figure 1.23 shows that with a PBG ground plane the insertion loss has been improved from 1.3 to 2.8 GHz, which is close to that of a conventional CPW [54].

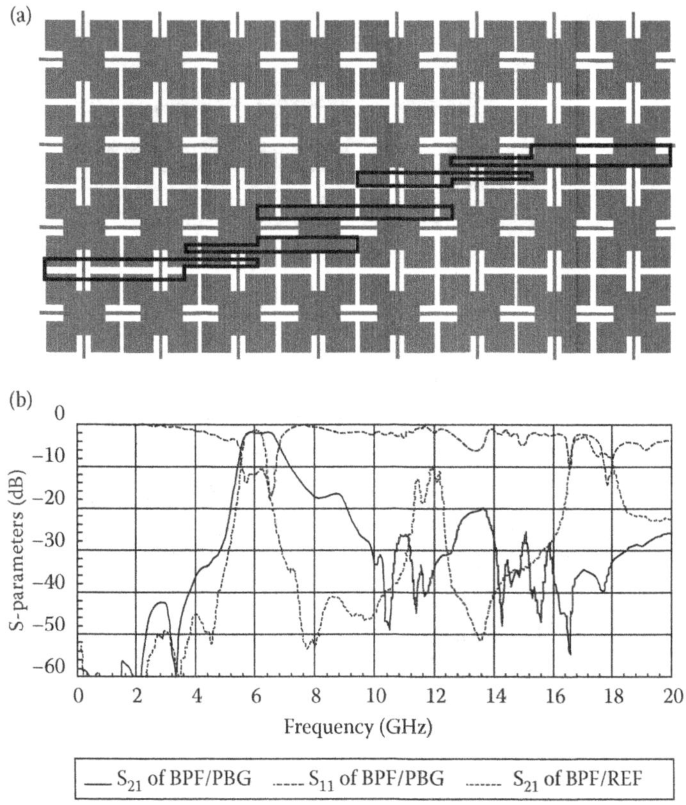

FIGURE 1.22
Parallel-coupled microstrip BPF with UC-PBG ground plane: (a) schematic diagram and (b) results compared with the conventional BPF. (F.-R. Yang et al., A uniplanar compact photonic-bandgap (UC-PBG) structure and its applications for microwave circuit, *IEEE Trans. Microw. Theory Tech.*, vol. 47, no. 8, pp. 1509–1514, Aug. © 1999 IEEE.)

A bidirectional antenna can be converted into a unidirectional antenna by placing the antenna over a reflector (conducting metal plate). The reflector needs to be placed at a distance of a quarter wavelength from the antenna to ensure that the reflected wave will be in phase with the direct wave in the forward direction. This results in constructive interference that will increase the antenna gain. In this case, the overall height of the antenna is great. If the antenna is placed too close to the reflector, it will result in destructive interference due to out-of-phase image currents in the reflector, leading to poor antenna performance. The unusual in-phase reflection property of EBG allows antennas to be placed close to the EBG reflector, thereby resulting in a low-profile antenna.

Figure 1.24 shows one horizontal antenna on a flat conducting metal sheet and one on a high-impedance surface. Based on the results shown in Figure 1.24c, the antenna on the metal plate reflects most of the energy, whereas the antenna on the high-impedance ground plane reflects only some power in the bandgap region. To get this same response with the metal sheet, the distance between the metal sheet and the antenna has to be a quarter wavelength. Thus, the high-impedance surface makes the antenna low-profile [11].

A low-profile wire antenna is demonstrated by using an EBG ground plane in References 55 and 56. Slot antennas with a backing cavity prevent backside radiation, which results in a unidirectional antenna. The cavity height needs to be an odd multiple of

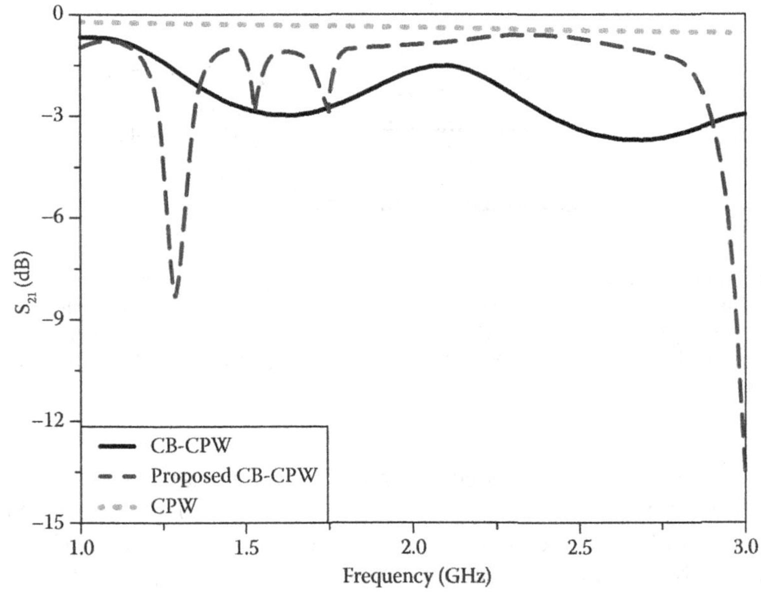

FIGURE 1.23
Comparison of transmission in CB-CPW, CB-CPW with PBG ground plane, and CPW. (Adapted from J. Wang et al., Design and application of a novel CB-CPW structure, *Prog. Electromagn. Res. M*, vol. 4, pp. 133–142, 2008.)

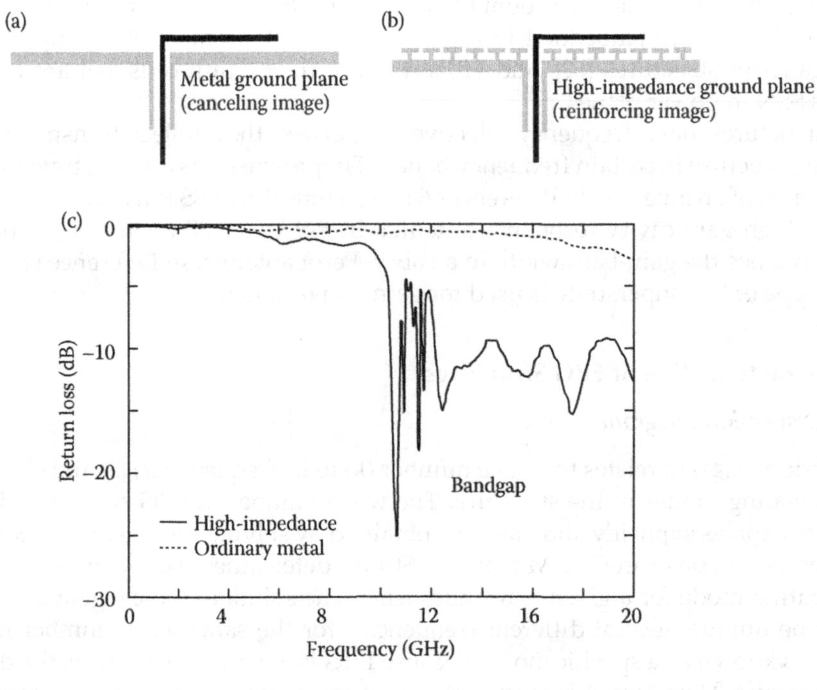

FIGURE 1.24
Horizontal wire antenna over (a) conducting flat metal plate, (b) high-impedance ground plane, and (c) return loss versus frequency characteristics. (D. Sievenpiper et al., Highimpedance electromagnetic surfaces with a forbidden frequency band, *IEEE Trans. Microw. Theory Tech.*, vol. 47, no. 11, pp. 2059–2074, Nov. © 1999 IEEE.)

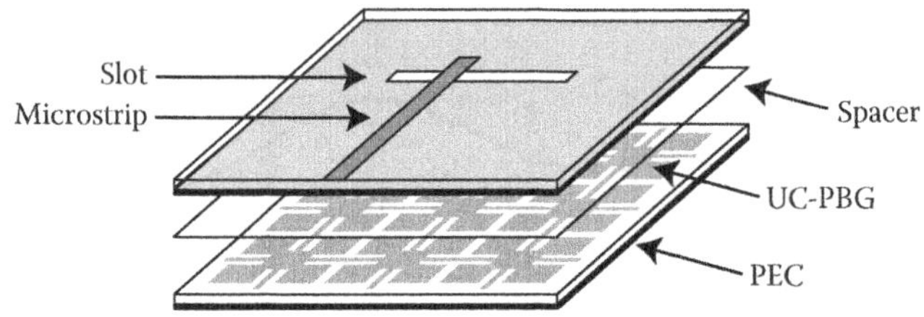

FIGURE 1.25
Schematic diagram of slot antenna backed with UC-PBG. (F. Yang, Y. Qian, and T. Itoh, Low-profile cavity-backed slot antenna using UC-PBG substrate, in *IEEE Antennas and Propagation Society International Symposium*, Salt Lake City, vol. 3, pp. 1796–1799, July 16–21 © 2000 IEEE; Adapted from C. Chang, Y. Qian, and T. Itoh, Analysis and applications of uniplanar compact photonic bandgap structures, *Prog. Electromagn. Res.*, vol. 41, pp. 211–235, 2003.)

the quarter wavelength to allow impedance matching. The quarter wavelength transforms the backing conductor into an open circuit at the slot antenna. Thus, the in-phase reflection or PMC property of PBG is equivalent to the open circuit at the surface. Therefore, using a PBG structure as the backing to a slot antenna with a thin substrate as the spacer between them can reflect the back radiations forward with a shorter cavity height, resulting in a low-profile antenna [57,58]. The schematic of a cavity-backed slot antenna using a UC-PBG structure is shown in Figure 1.25.

In high-speed packages, a simultaneous switching noise (SSN) is the most common noise that results in electromagnetic interference. Miniaturized EBG structures with high-k dielectrics [59], multi-via mushroom EBG structures [60], a double-surface EBG structure with one EBG surface embedded in a power plane [61], an alternating-impedance EBG with double-slit structures [62], and Hilbert curve fractal unit cells [63] are reported to suppress SSN in power planes.

EBG structures have frequency-selective properties; they reflect/transmit the signal across the structure in certain frequency bands. They are used as a superstrate for the gain enhancement of an antenna. In References 64–66, a reflecting FSS is used as the superstrate to form a high-gain cavity resonance antenna. In Reference 67, another reflecting FSS is used to enhance the gain-bandwidth in a Fabry–Perot antenna. In Reference 68, the transmission type of FSS superstrate is used for gain enhancement.

1.1.6 Characterization of EBG Structures

1.1.6.1 Dispersion Diagram

A dispersion diagram relates the wave number (k) to its frequency (f) [3] and characterizes the propagating modes in the structure. The wave number in EBG structures is usually difficult to express explicitly and has to be obtained by solving the eigenvalue equation [1]. The Eigenmode Solver in CST Microwave Studio determines the resonant frequency of a propagating mode for a given wave number k. The solution of the eigenvalue equation may not be unique; several different frequencies for the same wave number may exist. Each one is known as a specific mode. Because EBG is a periodic structure, the dispersion is also periodic. Therefore, it is sufficient to plot the dispersion within one single period, which is known as a Brillouin zone.

The Brillouin zone depends on the lattice arrangement of the EBG. When the periodic elements are arranged in a square lattice, then the Brillouin zone is square. If the unit cell

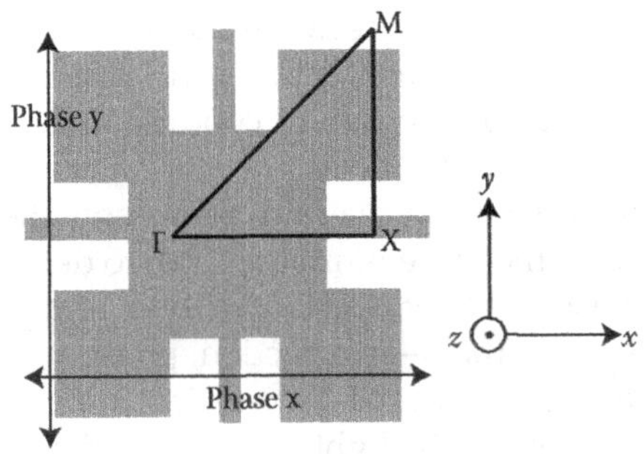

FIGURE 1.26
Unit cell of a UC-EBG structure showing irreducible Brillouin zone. (R. Coccioli et al., Aperture-coupled patch antenna on UC-PBG substrate, *IEEE Trans. Microw. Theory Tech.*, vol. 47, no. 11, pp. 2123–2130, Nov. © 1999 IEEE.)

is symmetric (as shown in Figure 1.26), then the computational region can be reduced to an irreducible Brillouin zone, which is represented by a triangle (Γ–X–M–Γ) as shown in Figure 1.26, where

$$\Gamma: k_x = 0, k_y = 0, \quad X: k_x = \frac{\pi}{p}, k_y = 0, \quad M: k_x = \frac{\pi}{p}, k_y = \frac{\pi}{p}$$

k_x, k_y are wave numbers in the x and y directions, respectively, and p is period.

To calculate the dispersion diagram of the EBG structure, it is sufficient to calculate along the irreducible triangle. The other triangular sections in the square Brillouin zone will exhibit the same behavior due to the symmetry of the structure.

The Eigenmode Solver of CST Microwave Studio is used to compute the dispersion diagram. The simulation setup is shown in Figure 1.27. Periodic boundaries are applied to the sides of the unit cell in the x and y directions, with a phase difference of phase x between

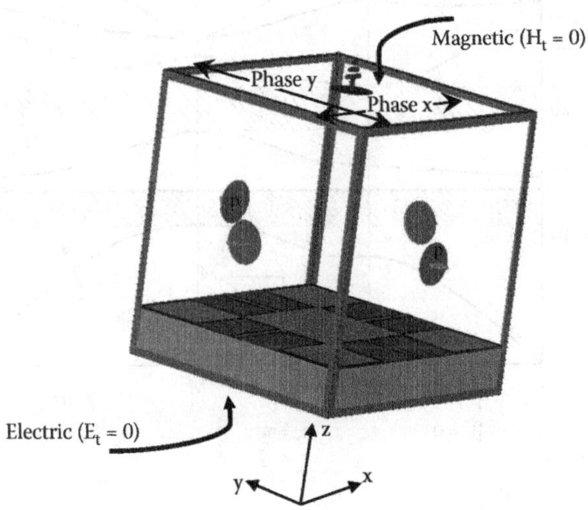

FIGURE 1.27
Simulation setup for dispersion diagram.

the walls in the x direction and phase y between the walls in the y direction. A magnetic boundary condition ($H_t = 0$) is applied at five times the substrate height above the top surface of the geometry, and an electric boundary condition ($E_t = 0$) is applied to the bottom side of the unit cell.

The horizontal axis in the dispersion diagram represents the wave number. The phase between walls is analogous to the wave number. Thus, to determine the eigenfrequencies along the Γ-X direction, phase y is kept constant at zero, and phase x is varied from 0° to 180°. For eigenfrequencies along the X–M direction, phase x is kept constant at 180°, and phase y is varied from 0° to 180°. For the M–Γ direction, both phase x and phase y are varied simultaneously from 180° to 0°. The light lines in the dispersion diagram indicate the frequencies of the waves propagating in free space for the corresponding wave number/phase. The frequency of the light lines are given by f = (phase*C/360*p) in the Γ–X region.

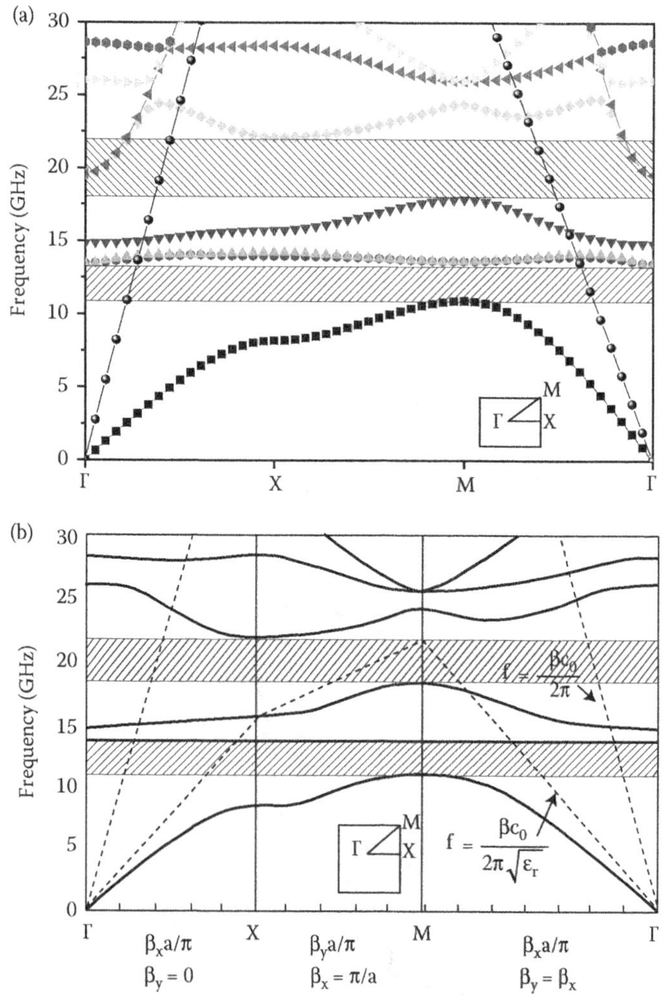

FIGURE 1.28
Dispersion diagram of UC-PBG structure with 25-mil substrate: (a) computed and (b) reported. (R. Coccioli et al., Aperture-coupled patch antenna on UC-PBG substrate, *IEEE Trans. Microw. Theory Tech.*, vol. 47, no. 11, pp. 2123–2130, Nov. © 1999 IEEE.)

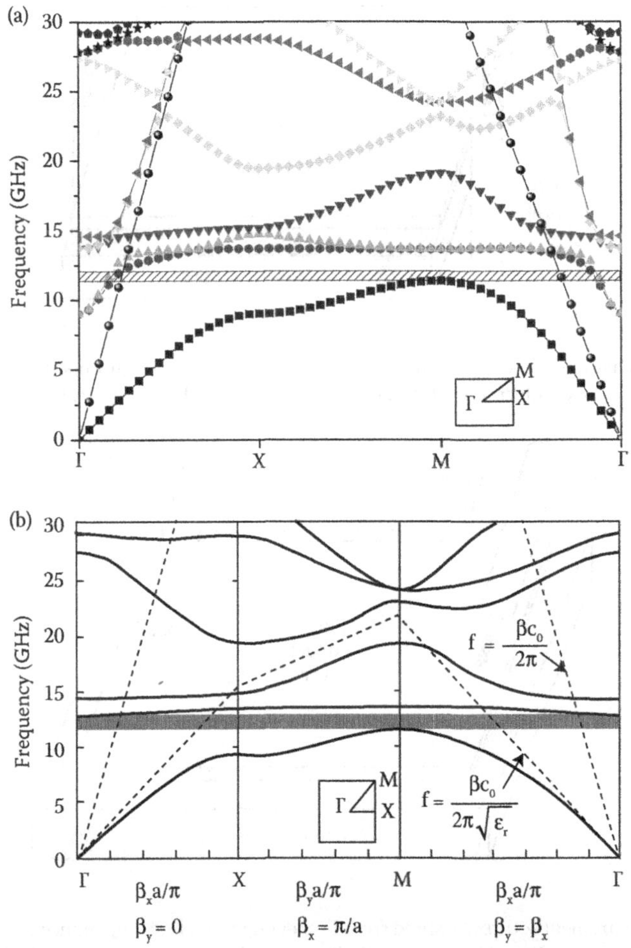

FIGURE 1.29
Dispersion diagram of UC-PBG structure with 50-mil substrate: (a) computed, (b) reported. (R. Coccioli et al., Aperture-coupled patch antenna on UC-PBG substrate, *IEEE Trans. Microw. Theory Tech.*, vol. 47, no. 11, pp. 2123–2130, Nov. © 1999 IEEE.)

In the M–Γ region, light-line frequencies are given by $f = (\text{phase} * C/360 * \sqrt{2}p)$, where C is the velocity of light in free space, and p is the period. The combined result of the three regions of the irreducible Brillouin zone and the light lines are plotted for the dispersion diagram.

To verify the CST simulation setup, the dispersion diagram is computed for the reported structure [14] shown in Figure 1.26, with substrate thicknesses of 25 and 50 mil; the computed and reported dispersion diagrams for these substrates are shown in Figures 1.28 and 1.29. The computed and reported dispersion diagrams match well between the light lines, which is required for the bandgap. In Figure 1.28, the space between the highest frequency of mode 1 and lowest frequency of mode 2, which falls beneath the light lines, is considered to be a bandgap. Similarly, around 20 GHz, there is gap between two modes where a propagating mode does not exist below the light lines; this is considered to be as a second bandgap.

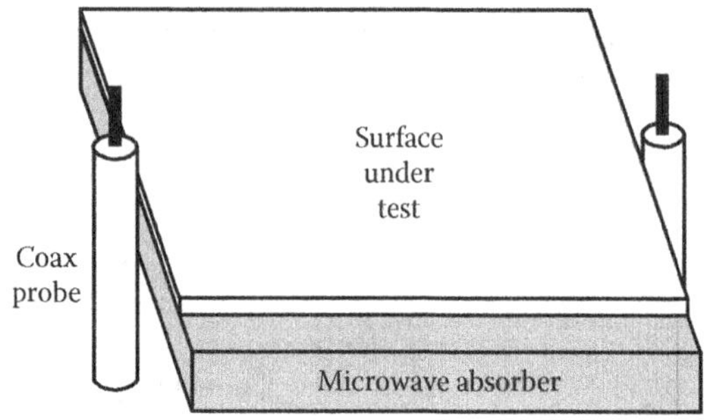

FIGURE 1.30
TM surface wave measurement setup. (Adapted from D. Sievenpiper, High-impedance electromagnetic surfaces, PhD dissertation. Dept. Elect. Eng., University of California at Los Angeles, Los Angeles, 1999.)

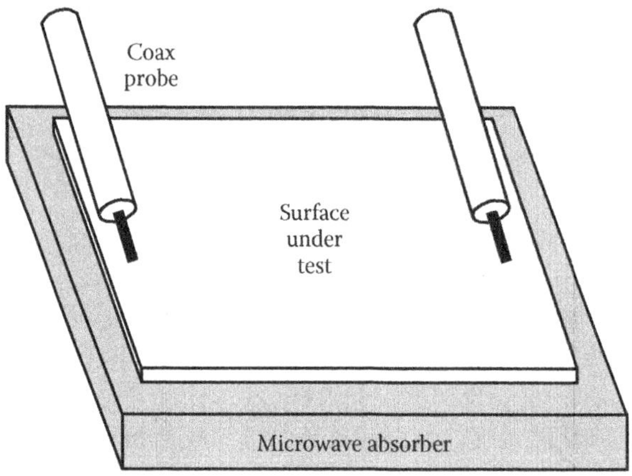

FIGURE 1.31
TE surface wave measurement setup. (Adapted from D. Sievenpiper, High-impedance electromagnetic surfaces, PhD dissertation. Dept. Elect. Eng., University of California at Los Angeles, Los Angeles, 1999.)

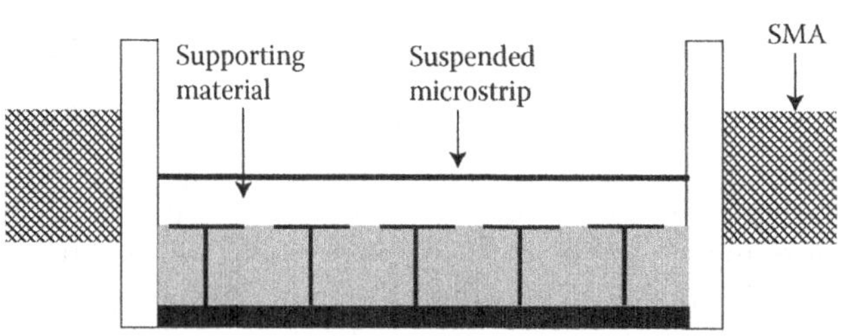

FIGURE 1.32
Method of suspended microstrip. (L. Yang et al., A novel compact electromagnetic-bandgap (EBG) structure and its applications for microwave circuits, *IEEE Trans. Microw. Theory Tech.*, vol. 53, no. 1, pp. 183–190, Jan. © 2005 IEEE.)

1.1.6.2 Surface Wave Measurement

Sievenpiper et al. [11,12] proposed methods to measure the surface properties of EBG structures. The transmission between two simple monopole probes positioned near the surface detects the presence of surface wave modes.

In transverse magnetic (TM) surface waves, the electric field forms loops that extend vertically out of the surface. A small monopole antenna placed normally with respect to the surface can measure the TM surface wave because the electric field of the probe and electric field of the TM surface wave are vertical [12]. The measurement setup is shown in Figure 1.30.

In transverse electric (TE) surface waves, the electric field is parallel to the surface, and the magnetic field forms vertical loops that protrude from the surface. The probes are placed parallel to the sheet to couple these waves [12]. The TE surface wave measurement setup is shown in Figure 1.31.

By placing the EBG structure as a test surface in the above measurement setup, the transmission between the antennas can predict the bandgap of the EBG structure.

1.1.6.3 Suspended Microstrip Method

The suspended microstrip method, proposed by M.Y. Fan, can be used for the bandgap characterization of EBG structures [27,35]. The microstrip line fabricated on a thin substrate is placed on an EBG structure, as shown in Figure 1.32. The transmission characteristics can determine the bandgap characteristics of the EBG structure.

1.1.6.4 Truncated Microstrip Line

A truncated microstrip line coupled to an EBG structure with a small gap at both ends is used to measure bandgaps [14,29], as shown in Figure 1.33. The bandgap can be determined by the transmission characteristics.

1.1.6.5 Reflection Phase Characteristics

The EBG structure acts as a PMC surface, providing in-phase reflection. Figure 1.34 shows the setup to measure the reflection phase characteristics. The two horns are placed next to each other, facing the test surface. First, the transmission phase between two antennas is measured with a metal plate and then with an EBG surface. The reflection phase

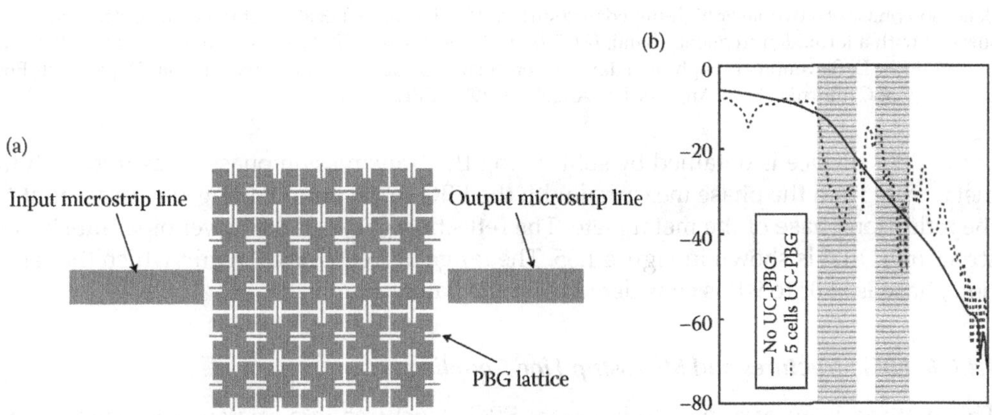

FIGURE 1.33
Transmission through UC-PBG lattice by truncated microstrip line: (a) top view and (b) transmission characteristics. (R. Coccioli et al., Aperture-coupled patch antenna on UC-PBG substrate, *IEEE Trans. Microw. Theory Tech.*, vol. 47, no. 11, pp. 2123–2130, Nov. © 1999 IEEE.)

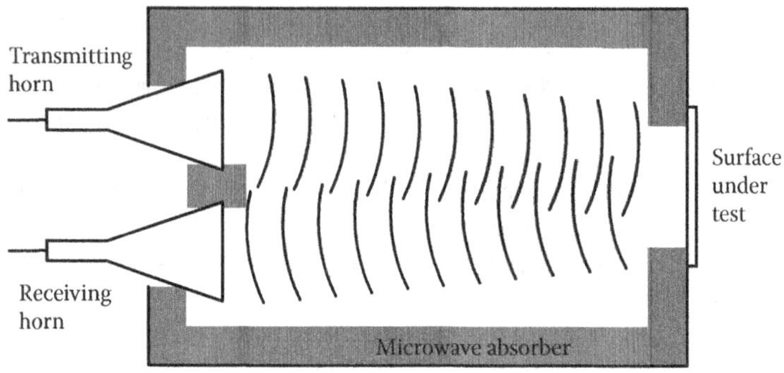

FIGURE 1.34
Reflection phase measurement setup. (D. Sievenpiper et al., Highimpedance electromagnetic surfaces with a forbidden frequency band, *IEEE Trans. Microw. Theory Tech.*, vol. 47, no. 11, pp. 2059–2074, Nov. © 1999 IEEE; D. Sievenpiper, High-impedance electromagnetic surfaces, PhD dissertation. Dept. Elect. Eng., University of California at Los Angeles, Los Angeles © 1999 IEEE.)

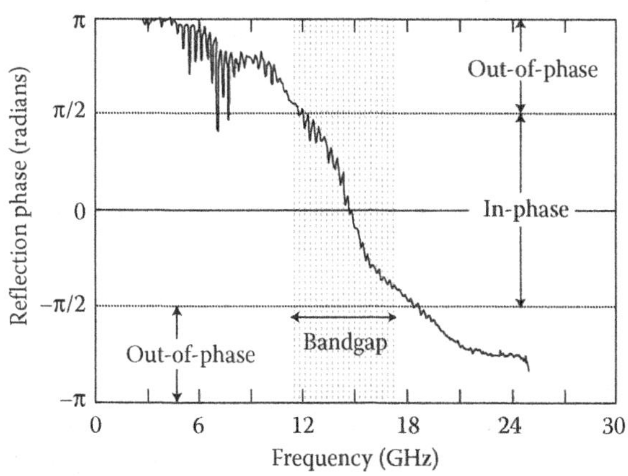

FIGURE 1.35
Reflection phase of a two-layer high-impedance surface. (D. Sievenpiper et al., Highimpedance electromagnetic surfaces with a forbidden frequency band, *IEEE Trans. Microw. Theory Tech.*, vol. 47, no. 11, pp. 2059–2074, Nov. © 1999 IEEE; D. Sievenpiper, High-impedance electromagnetic surfaces, PhD dissertation. Dept. Elect. Eng., University of California at Los Angeles, Los Angeles © 1999 IEEE.)

of the EBG surface is obtained by subtracting the transmission phase measured with the metal plate from the phase measured with the EBG surface and adding 180° to account for the reflection phase of the metal plate. The reflection phase of the Sievenpiper mushroom structure [11,12] is shown in Figure 1.35. The range of frequencies within which the reflection phase is –90° to 90° is considered to be the bandgap.

1.1.6.6 EBG Structures and Microstrip Line Coupling

This method is an easy way to integrate EBG structures into microwave integrated circuits. In References 69–73, mushroom EBG structures are coupled to a microstrip line to achieve notch-filter characteristics using the bandgap property of the EBG structure.

Figure 1.36 shows the top and cross-sectional views of a mushroom EBG structure coupled to a transmission line. A similar type of arrangement for a notch filter is demonstrated in

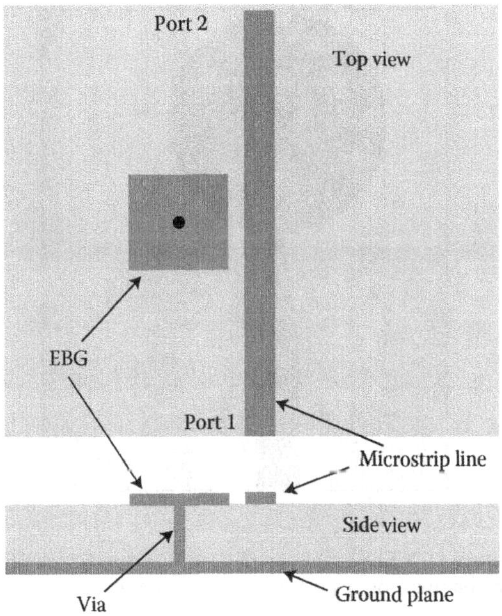

FIGURE 1.36
Mushroom EBG structure coupled to a microstrip transmission line. (L. Peng and C.-L. Ruan, UWB band-notched monopole antenna design using electromagneticbandgap structures, *IEEE Trans. Microw. Theory Tech.*, vol. 59, no. 4, pp. 1074–1081, Apr. © 2011 IEEE.)

Reference 74 with a simplified composite right/left-handed resonator and in Reference 75 with a folded stepped-impedance resonator (SIR). All of the previously mentioned structures include vias connecting the structure to the ground. Structures with vias can be difficult to fabricate. In Reference 76, split ring resonators (SRRs) are coupled to a microstrip line to obtain notch-filter characteristics. Recently, an interdigital capacitance loading loop resonator (IDCLLR) was coupled to a microstrip transmission line to achieve a notch filter [77]. The planar IDCLLR structure coupled to a microstrip line is shown in Figure 1.37.

In this book, a planar EBG structure without vias is coupled to a microstrip transmission line in order to achieve notch-filter/BSF characteristics by utilizing the bandgap property of the EBG structure.

1.1.7 Reconfigurable EBG Structures

One of the limiting factors in using an EBG structure is the available bandwidth. The useable frequency range of an EBG structure can be extended through electrical tuning/reconfiguration using lumped elements. The equivalent circuit model of an EBG structure is basically an LC circuit. The resonant frequency, reflection phase, and slow-wave factor depends on L and C. Electrical tuning of an EBG structure can be achieved by using a variable capacitor (i.e., the varactor diode in the structure).

A proximity feed microstrip antenna is reconfigured by placing the feed line on top of an EBG substrate. The mushroom EBG structure in which vias are connected to the ground plane through diode switches is used. By switching diodes between two states, on and off, loading to the microstrip line is modified, and the frequency reconfiguration of the patch antenna is achieved in Reference 78.

FIGURE 1.37
Band-notched UWB filter using an IDCLLR structure. (T. Li et al., Compact UWB band-notched antenna design using interdigital capacitance loading loop resonator, *IEEE Antennas Wirel. Propag. Lett.*, vol. 11, pp. 724–727, June © 2012 IEEE.)

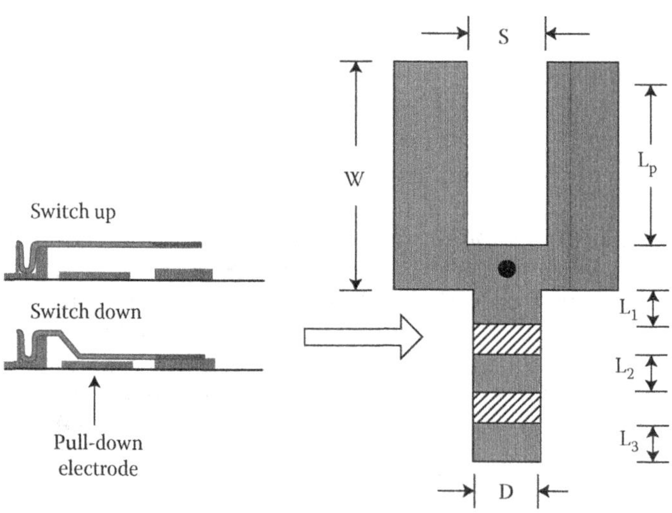

FIGURE 1.38
Reconfigurable fork-like EBG unit cell. (L. Yang et al., A novel compact electromagnetic-bandgap (EBG) structure and its applications for microwave circuits, *IEEE Trans. Microw. Theory Tech.*, vol. 53, no. 1, pp. 183–190, Jan. © 2005 IEEE.)

A filter that can reconfigure bandpass to bandstop characteristics at the same frequency in a CPW configuration using an EBG structure is demonstrated in Reference 79. A micro-electromechanical systems (MEMS)-based tunable BSF using an EBG structure is proposed in Reference 80. An independently tunable dual-band EBG AMC surface is presented in Reference 81 by using concentric square loops in the unit cell. A multilayer independently tunable dual-band EBG AMC is presented in Reference 82. Beam steering using reconfigurable periodic structures in the ground plane is demonstrated in Reference 83. A tunable EBG surface is achieved by connecting an array of square metallic patches with tunable capacitors in References 84–86. Varactors are placed between adjacent pads in mushroom EBG structures to achieve tunable reflection-phase characteristics and reflective beam steering, as demonstrated in Reference 87.

The bandgap of the fork-like mushroom EBG structure is reconfigured by using MEMS switches placed at the root of each stretched strip of the structure [27]. The fork-like EBG unit cell with a reconfigurable configuration is shown in Figure 1.38. By switching MEMS switches, the length of the stretched strip can be varied, thereby achieving reconfigurability.

Compact reconfigurable planar EBG structures based on short-circuited hairpin resonators are demonstrated in References 88 and 89. The EBG unit cell is shown in Figure 1.39a. Three unit cells are placed in between the microstrip line and the ground plane to create a BSF based on the bandgap property of the EBG structure. Its top and side views are shown in Figure 1.39b. In order to reconfigure the unit cell, the capacitance of the unit cell is varied by placing a varactor in between the arms of the unit cell (as shown in the inset

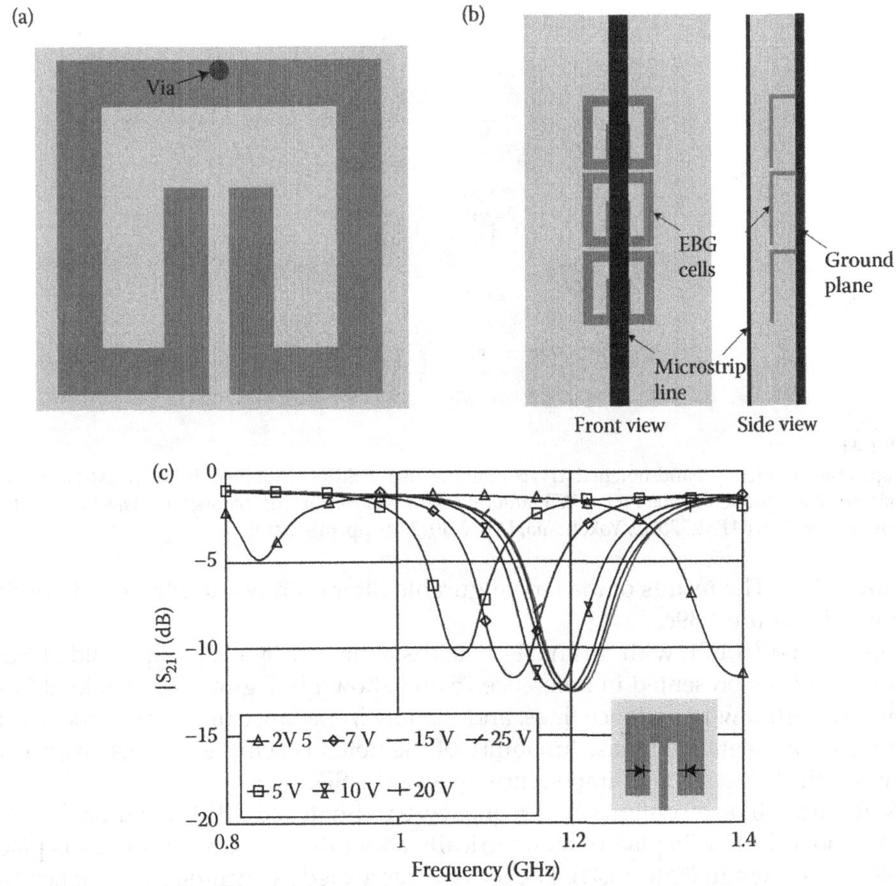

FIGURE 1.39
Short-circuited hairpin resonator: (a) schematic diagram of unit cell, (b) cells loaded to a transmission line, and (c) result of reconfigurable filter. (Adapted from L. Inclán-Sánchez et al., Reconfigurable unit cell for EBG and artificial surfaces based on hairpin resonators, in *IEEE MTT-S International Microwave Optoelectronics Conference (IMOC)*, Natal, Brazil, Oct. 29–Nov. 1, 2011, pp. 191–195; Adapted from O. Quevedo-Teruel et al., *IEEE Microw. Wirel. Compon. Lett.*, vol. 23, no. 9, pp. 462–464, Sep. 2013.)

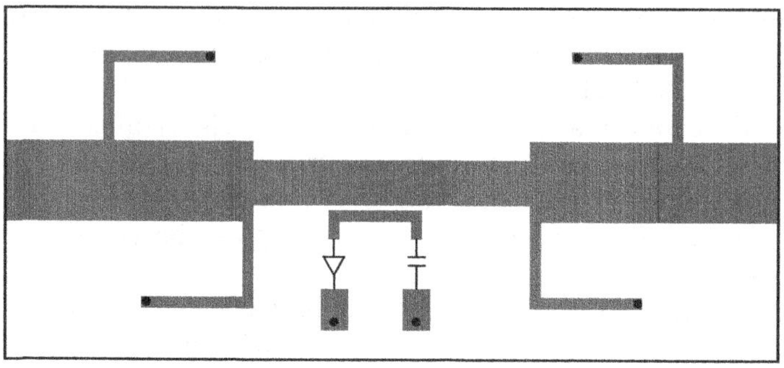

FIGURE 1.40
UWB filter with tunable notch based on folded SIR. (Adapted from F. Wei et al., *Electron. Lett.*, vol. 47, no. 22, pp. 1229–1230, Oct. 2011.)

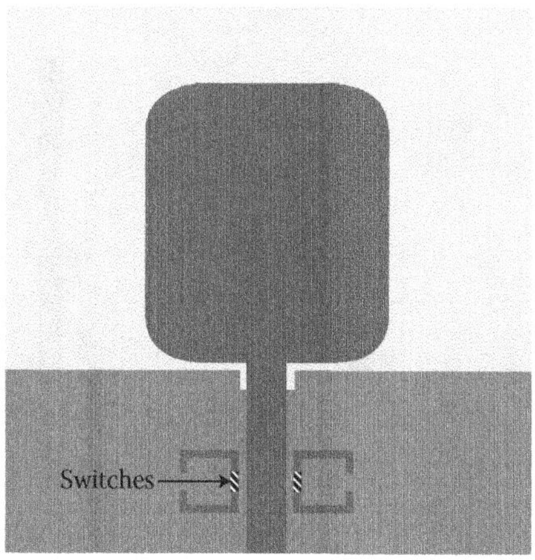

FIGURE 1.41
Reconfigurable frequency band-notched UWB antenna using SRRs. (Adapted from M. Al-Husseini et al., A reconfigurable frequency-notched UWB antenna with split-ring resonators, in *Asia-Pacific Microwave Conference Proceedings (APMC 2010)*, Yokohama, Dec. 7–10, 2010, pp. 618–621.)

of Figure 1.39c). The results of the reconfigurable filter with two diodes, one in each arm, are shown in Figure 1.39c.

A compact UWB BPF with a tunable notch is achieved by coupling a folded SIR to a microstrip line, as presented in Reference 75 and shown in Figure 1.40. A folded SIR consists of low-high-low impedance lines, and the low-impedance lines are connected to the ground through metalized vias. Tunability of the notch is achieved by placing a varactor in between the low- and high-impedance lines in the SIR.

In Reference 90, a reconfigurable-frequency band-notched UWB antenna is achieved using two identical SRRs placed symmetrically about the feed line. A switch is placed in each SRR (as shown in Figure 1.41). The SRRs are activated by turning the switches on, and a band notch appears. The notch disappears when the switches are turned off.

1.1.8 Resonant-Like EBG Structures

EBG structures were initially used in the optical domain (known then as PBG structures). The filtering effect of the periodic structures is due to the Bragg phenomenon. Bragg-like

EBG structures have a periodicity around $\lambda_g/2$ (λ_g is the guide wavelength in the medium) that results in a stopband at the frequency corresponding to λ_g. The structures previously used in the optical domain are scaled down to microwave frequencies. Along with the Bragg phenomenon, an additional phenomenon of the unit cell's self-resonance should be considered at microwave frequencies. Apparently, this phenomenon does not appear at optical frequencies, as the self-resonant effects of the unit cells are negligible in comparison to the filtering response of the whole periodic structure. Hence, the contribution of each individual unit cell to the overall periodic filtering effect is very small. However, the self-resonances cannot be neglected at microwave frequencies; these resonances will produce a rejection band that is sharper and often deeper than that obtained from the EBG effect [91].

Caloz et al. [92,93] introduced an anisotropic UC-PBG, a microstrip structure with a ground plane consisting of an array of etched slots of alternating widths. A slot of alternating widths is considered to be a unit cell. An array of 7×7 unit cells are etched on the ground plane and depicted in Figure 1.42. When the transmission line is placed along the *z*-axis, the induced current flows freely through the structure and the signal is transmitted; thus, this direction is the propagation direction. When the transmission line is placed along the *y*-axis, the slots break the current and does not allow the signal to propagate along the line; thus, this direction is the attenuation direction, resulting in a bandgap that is not due to Bragg-like diffraction. In this case, the bandgap is due to the resonance of the resonator that is formed by the total number of cells in the transverse direction. The resonance occurs when the total length of the resonator is equal to a half wavelength. The number of cells along the transmission lines determines the depth of the bandgap but not the bandgap position.

Because the bandgap depends on the resonator formed by the periodic structures that are transverse to the transmission line with a periodicity much smaller than the wavelength, this type of EBG is a resonant-like EBG structure. It is worth noting that the characteristics of this EBG structure are not due to the Bragg effect.

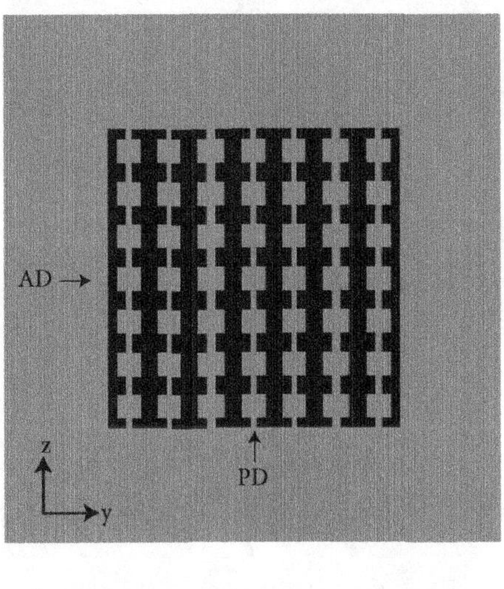

FIGURE 1.42
Anisotropic UC-PBG etched on the ground plane of microstrip line. (C. Caloz and T. Itoh, Multilayer and anisotropic planar compact PBG structures for microstrip applications, *IEEE Trans. Microw. Theory Tech.*, vol. 50, no. 9, pp. 2206–2208, Sep. © 2002 IEEE.)

In Reference 94, a resonator-like EBG structure for printed slot-line devices is presented that is more compact than Bragg-like EBG structures for printed slot-line devices. This structure, which is the dual version of the compact anisotropic UC-PBG structure [93], is obtained by etching the metallic line sections on the opposite side of the slot line.

Quevedo-Teruel et al. [89] propose a short-circuited hairpin resonator, which is a resonant-like EBG structure in which the dominant effect is the resonance related to the dimensions of the unit cell.

The EBG structure proposed in this book also has a dominant resonance effect that depends strongly on the geometry and dimensions of the unit cell.

2

Planar Periodic Structures as Electromagnetic Bandgap Surfaces

2.1 Introduction

The mushroom electromagnetic bandgap (EBG) structure, introduced by Sievenpiper et al. [11], and the uniplanar compact electromagnetic bandgap (UC-EBG) structure, introduced by Yang et al. [13], have made it convenient to use periodic structures at microwave frequencies. The UC-EBG structure in particular is generating significant interest among researchers because, without the via, it is easier to fabricate than the mushroom EBG structure. The UC-EBG structure is described in detail in Section 1.1.2. The resonance frequency is inversely proportional to the effective inductance and capacitance of the unit cell, as given in Equation 1.1. Increasing the effective inductance and capacitance of the unit cell without increasing the unit cell size causes the resonance frequency to decrease, thereby making the structure compact. Miniaturized planar EBG structures are reported in References 30–35. This chapter focuses on developing another compact planar EBG structure, its characterization, and its integration with microstrip circuits.

2.2 Proposed EBG Structure

The proposed planar EBG unit cell is shown in Figure 2.1. The size of the unit cell is 4.32 mm × 4.32 mm. A metal pattern is etched on top of a conductor backed GML 1000 substrate, with relative permittivity $\varepsilon_r = 3.2$ and thickness $h = 0.762$ mm. The metal pattern can be seen in black in Figure 2.1. This metal pattern is formed by appropriately modifying the conventional uniplanar EBG structure reported in References 13 and 14. For comparison, a conventional uniplanar EBG with the same substrate parameters as those of the proposed EBG structure is considered. Its dimensions are shown in Figure 2.2. In the proposed structure, the meander lines from the center of the structure to its four corners provide the inductance, and the capacitance is provided by the interdigital lines on the periphery of the unit cell. Comparing the proposed and conventional EBG structures with the same overall size (4.32 mm × 4.32 mm) in Figures 2.1 and 2.2 reveals that the length of the lines providing inductance and the length of the gap providing capacitance in the proposed structure are greater than those that exist in the conventional structure, resulting in higher overall inductance and capacitance. Thus, the resonance frequency shifts downward, making the proposed structure more miniaturized than the conventional structure.

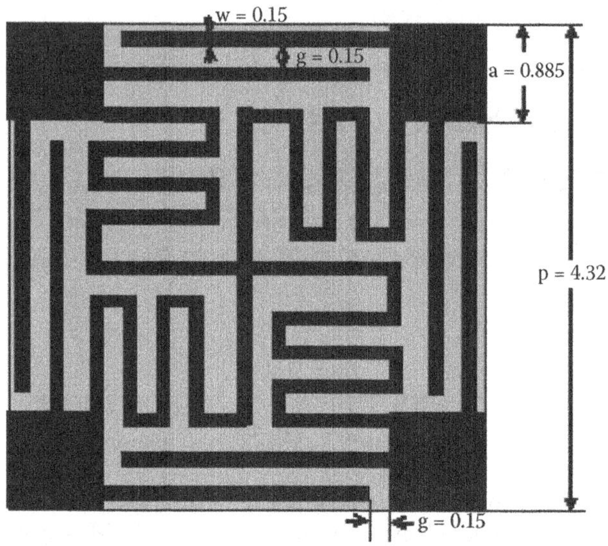

FIGURE 2.1
Proposed planar EBG unit cell.

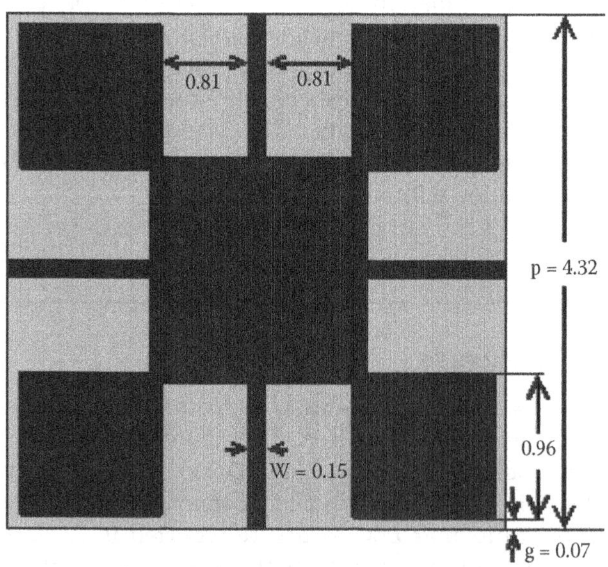

FIGURE 2.2
Conventional planar EBG unit cell.

2.3 Dispersion Diagrams

The dispersion diagrams of the proposed EBG structure (Figure 2.1) and the conventional planar EBG structure (Figure 2.2) are obtained using the procedure described in Section 1.1.6.1 and are shown in Figures 2.3 and 2.4, respectively. The dispersion diagram is computed along the irreducible Brillouin triangle Γ-X-M-Γ shown in the inset of the figures,

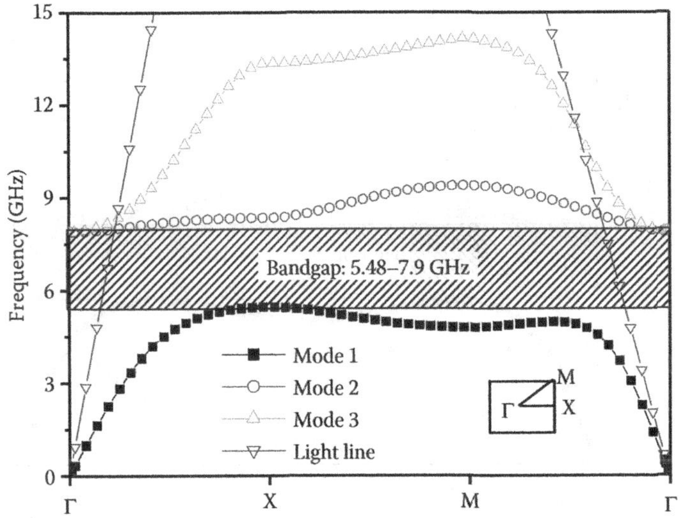

FIGURE 2.3
Dispersion diagram of the proposed EBG structure.

where Γ is $k_x = 0$ and $k_y = 0$; X is $k_x = \pi/p$ and $k_y = 0$; M is $k_x = \pi/p$ and $k_y = \pi/p$; k_x and k_y are wave numbers in x and y directions; and p is the period. It is seen from the dispersion diagrams that the proposed structure offers a 5.48–7.9 GHz bandgap. The bandgap of the conventional EBG structure (9.41–12.41 GHz) demonstrates that a nearly 38% reduction in the bandgap center frequency is achieved in the proposed structure. The fractional bandwidth is also increased in the proposed structure. Because the increase in inductance due to the meander lines is greater than the increase in capacitance due to the interdigital lines, the bandwidth of the stopband is increased.

As seen from the unit cell in Figure 2.1, the structure has rotational symmetry. Therefore, based on symmetry, it is appropriate to take the square Γ-X-M-Y-Γ as the irreducible

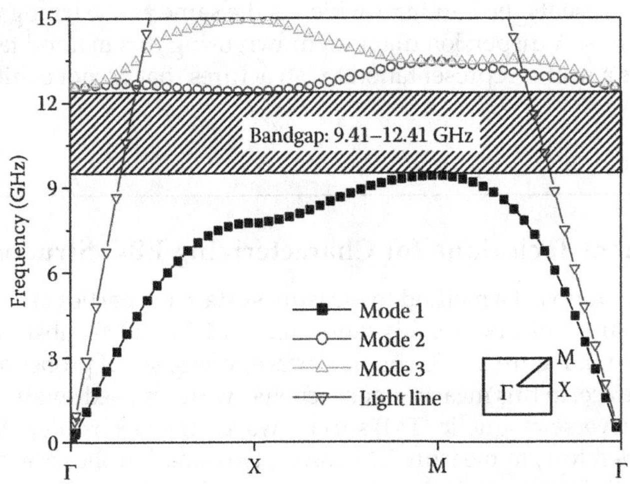

FIGURE 2.4
Dispersion diagram of the conventional planar EBG structure.

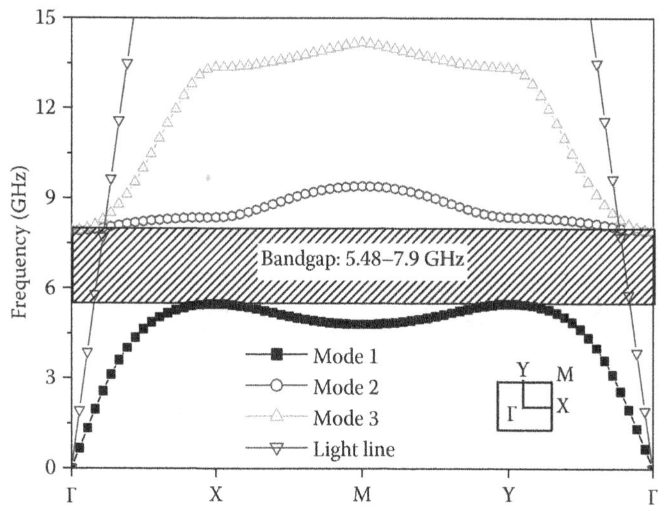

FIGURE 2.5
Dispersion diagram of the proposed EBG structure along the Γ-X-M-Y-Γ irreducible Brillouin square.

Brillouin zone, where Γ is $k_x = 0$ and $k_y = 0$; X is $k_x = \pi/p$ and $k_y = 0$; M is $k_x = \pi/p$ and $k_y = \pi/p$; Y is $k_x = 0$ and $k_y = \pi/p$; k_x and k_y are wave numbers in x and y directions; and p is the period. The dispersion diagram along this square is shown in Figure 2.5. It can be seen from the figure that the bandgap is the same whether the dispersion diagram is drawn along a triangular irreducible zone or a square irreducible zone.

The dispersion diagrams shown in Figure 2.5 are drawn along the irreducible zone. The dispersion diagram can also be drawn for the entire structure (arbitrary direction) by varying phase x from −180° to 180° at each value of phase y from −180° to 180°. Figure 2.6 shows a dispersion diagram drawn throughout the structure. The structure is simulated with the Eigenmode solver in the CST Microwave Studio software. The results are then used in MATLAB® to plot the points shown in Figure 2.6. The space between mode 1 and mode 2 clearly represents the bandgap, which is the same as the bandgap in the previous dispersion diagrams. A dispersion diagram drawn using this method takes more simulation time, but it is a useful representation for structures that do not exhibit symmetry.

2.4 Measurement Technique for Characterizing EBG Structure

Sievenpiper et al. proposed a method to measure surface properties of EBG structures [12]. To do so, a 7 × 7 array of EBG cells is fabricated on GML 1000 substrate. The fabricated structure is shown in Figure 2.7. To couple surface waves, small probes are placed close to the fabricated surface. In this measurement scheme, we have used small coaxial connectors as probes. In transverse magnetic (TM) surface waves, the electric field is in loops normal to the surface. Therefore, to measure TM waves, two small probes are placed close to the substrate with their center conductor also normal to the surface, as shown in Figure 1.30. For transverse electric (TE) surface waves, the magnetic field is in loops normal to the surface. Therefore, two probes are placed close to the surface with their center conductors

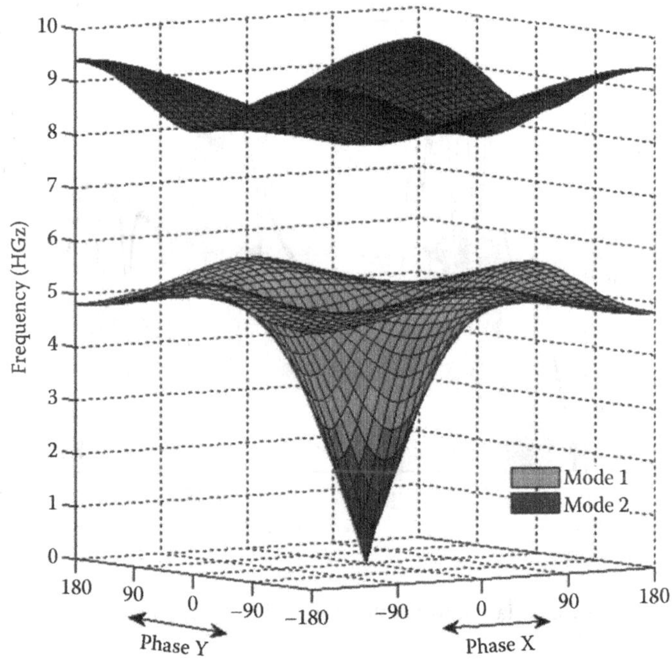

FIGURE 2.6
Dispersion diagram of the entire proposed EBG structure.

FIGURE 2.7
Photograph of fabricated array of 7 × 7 proposed EBG cells.

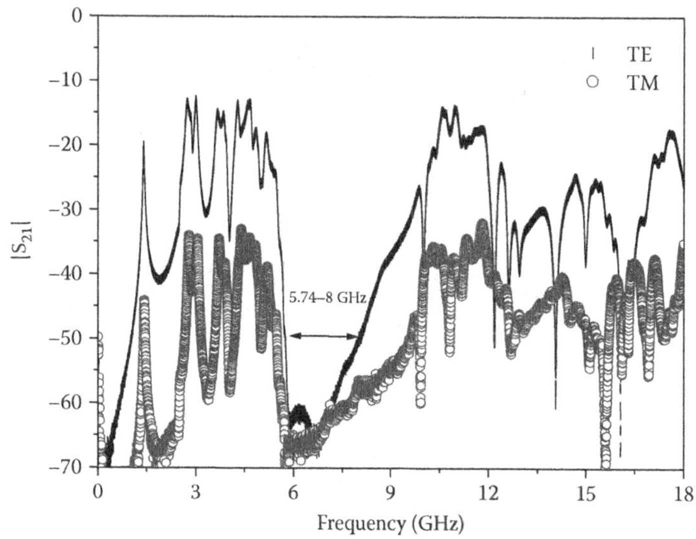

FIGURE 2.8
Measured results of TE and TM surface waves in the proposed EBG structure.

parallel to the surface, as shown in Figure 1.31. These two probes are connected to a ZVB20 Rohde & Schwarz vector network analyzer. The measured transmission characteristics of the two wave experiments are shown in Figure 2.8. It is observed that surface waves are suppressed below −60 dB in a frequency band of 5.74–8 GHz with the proposed structure. This closely matches the bandgap obtained in the dispersion diagram.

2.5 EBG Loaded Transmission Line

Another method to characterize the proposed EBG structure is capacitive coupling of the EBG cells to a transmission line. When a one-dimensional array of EBG unit cells is coupled to a 50 Ω microstrip transmission line through a small gap, the EBG structure is capacitively coupled to the transmission line. The EBG structure interacts with electromagnetic waves along the transmission line and prohibits wave propagation at a specific band due to its resonant nature. This band depends on the geometry of the unit cell (the inductance and capacitance) and the coupling capacitance between the transmission line and the EBG structure. The number of cells along the line influences the depth of the bandgap. In order to increase the depth, another one-dimensional array can be placed on the other side of the transmission line.

The schematic diagram of this arrangement is shown in Figure 2.9a. The spacing (d) between the line and the array is 0.1 mm. The circuit is fabricated on GML 1000 substrate, and its photograph is shown in Figure 2.9b. The transmission characteristics are shown in Figure 2.9c. As expected, due to the resonance of the EBG structure, a band-notch is seen in the transmission characteristics. The measured 3 dB bandgap is 4.95–5.37 GHz. This bandgap is smaller than the bandgap seen in the dispersion diagram. This is a result of the additional capacitance between the line and EBG structure and can also be explained by

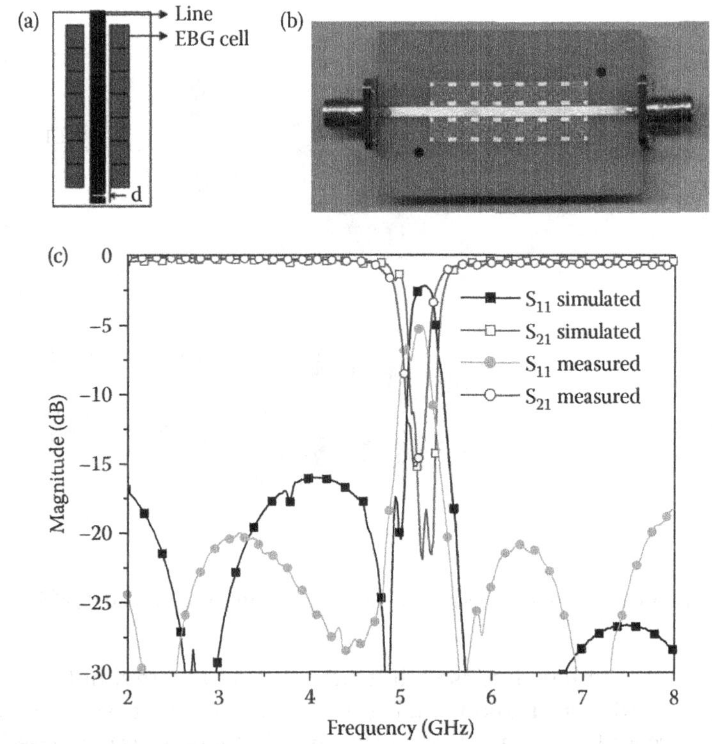

FIGURE 2.9
EBG structure coupled to a microstrip line. (a) Schematic diagram. (b) Photograph of fabricated circuit. (c) Simulated and measured results.

the fact that the dispersion diagram is computed for two-dimensional infinite structures. In the stopband, even though S_{21} is −15 dB, S_{11} is not close to 0 dB. Instead, it is up to −5 dB, which demonstrates there is some amount of energy confined within the resonant structures and the filter is not reflecting the signal completely in the stopband. This bandgap is due to self-resonance of the EBG structure rather than the Bragg effect [89]. The advantage at microwave frequencies is that the resonance effects are dominant and cannot be neglected as they can be at optical frequencies (PBG). This makes the EBG structure compact, whereas for bandgap due to Bragg effect, the unit cells should be at least 0.5 λ_g long along the transmission direction.

Figure 2.10 shows the surface current plots, taken from CST Microwave Studio at 5.25 GHz in the stopband and at 6.1 GHz in the passband. At 6.1 GHz (Figure 2.10a), the current is passing from port 1 to port 2. At 5.25 GHz (Figure 2.10b), when port 1 (right side) is excited, the first few cells from port 1 show higher surface current density than the later cells. This indicates that, at the cells' resonance frequency, energy is stored and not allowed to propagate, resulting in the bandgap at 5.25 GHz.

To study the effects of the meander line and interdigital capacitors in the circuit, three structures are analyzed. The proposed unit cell shown in Figure 2.1 is called structure A. In structure A, the meander line is present and the gap (g) between interdigital lines is 0.15 mm. In structure B (Figure 2.11), the meander line is replaced with straight lines, and the gap between interdigital lines is 0.15 mm. Structure C is similar to A, with g = 0.1 mm. Results of a one-dimensional array of structure A loaded to a 50 Ω transmission line are

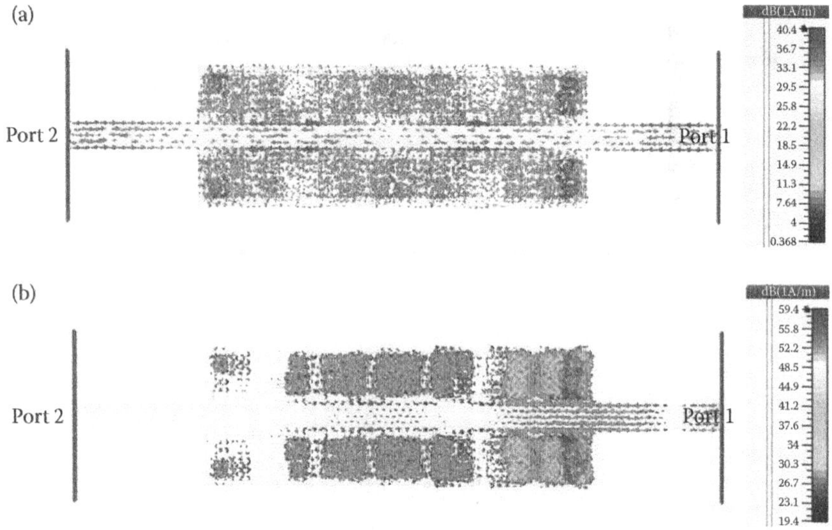

FIGURE 2.10
Surface current plot of EBG loaded transmission line (a) at 6.1 GHz in passband and (b) at 5.25 GHz in stopband.

shown in Figure 2.9c. Similar circuits with structures B and C are simulated, and a comparison of the three simulations is shown in Figure 2.12. Compared to structure A, structure B has less inductance and the same capacitance. Since inductance is decreased, the bandgap shifts to a higher frequency for B. Structure C's inductance is the same as structure A's, but the capacitance is higher, which results in the bandgap shifting to a lower frequency. The effects of the meander line and interdigital capacitors are summarized in Table 2.1.

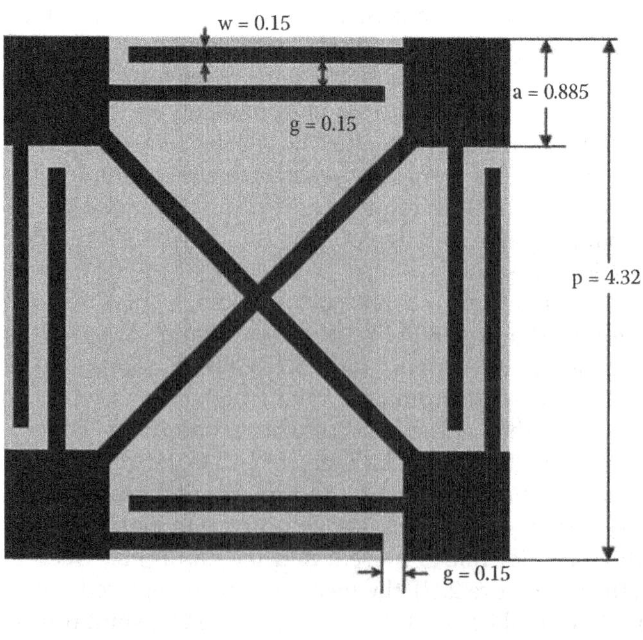

FIGURE 2.11
Unit cell of structure B. (All dimensions are in mm.)

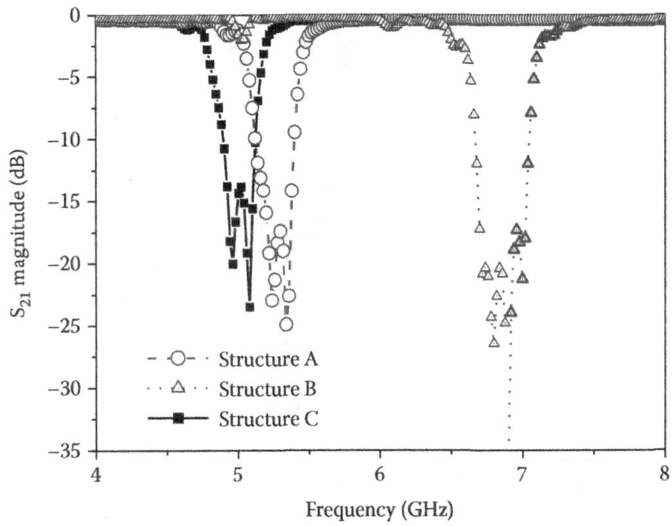

FIGURE 2.12
Simulated results of various proposed structures loading a transmission line.

TABLE 2.1
Summary of Results of Various Proposed Structures Loading a Transmission Line

Structure	Meander Line	g (mm)	Center Frequency (GHz)	10 dB-Bandwidth (GHz)
A	Yes	0.15	5.26	0.28
B	No	0.15	6.86	0.39
C	Yes	0.1	5.01	0.23

The circuit shown in Figure 2.9b is simulated with variation in spacing (d) between the transmission line and the EBG structures. The simulated results are shown in Figure 2.13. As d decreases, coupling to the structure is increased. Thus, the depth of the bandgap rejection is improved.

The proposed structure is also compared with the conventional structure (Figure 2.2) using this method of loading the EBG to a microstrip transmission line. The simulated results comparing the proposed structure (Figure 2.1) and the conventional structure (Figure 2.2) are shown in Figure 2.14. The bandgap obtained with the conventional structure is 12.61–13.67 GHz. It is clearly evident that the bandgap for the proposed structure is smaller than that of the conventional structure. This method, too, verifies that the proposed structure is compact.

2.6 Equivalent Circuit Model

This section models the equivalent circuit of the proposed EBG structure coupled to a transmission line. Figure 2.15 shows the metal pattern of the unit cell (Figure 2.1) coupled to a microstrip transmission line. This metal pattern is on top of the grounded substrate

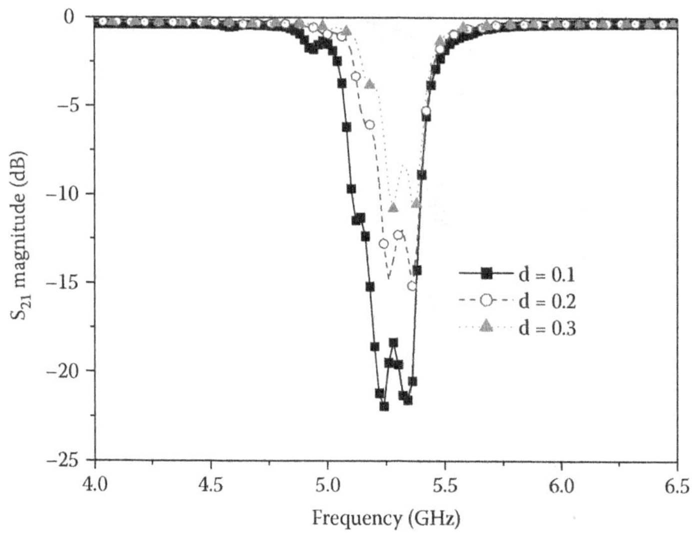

FIGURE 2.13
Simulated results of the proposed EBG structure loading a transmission line for various values of d.

($\varepsilon_r = 3.2$ and $h = 0.762$ mm). The unit cell consists of four small pads, four meander lines extending from the pads and connected at the center, and two-finger interdigital capacitors that are connected between the pads. The equivalent circuit of the unit cell is obtained by combining the circuit model of an inductor, an interdigital capacitor, and the coupling capacitance that represents the coupling between the unit cell and the transmission line. The metal pattern of the unit cell's meander line connected to the pad is shown in Figure 2.16a.

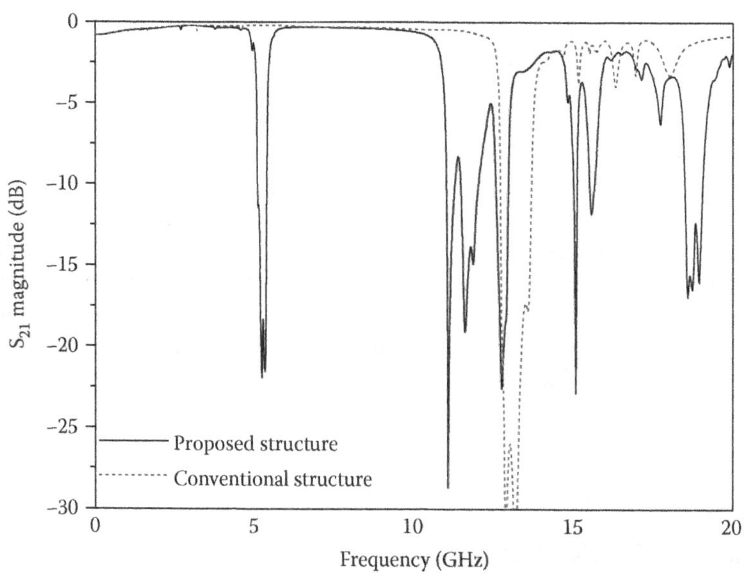

FIGURE 2.14
Simulated results comparing the proposed structure (Figure 2.1) and conventional structure (Figure 2.2) loading a transmission line.

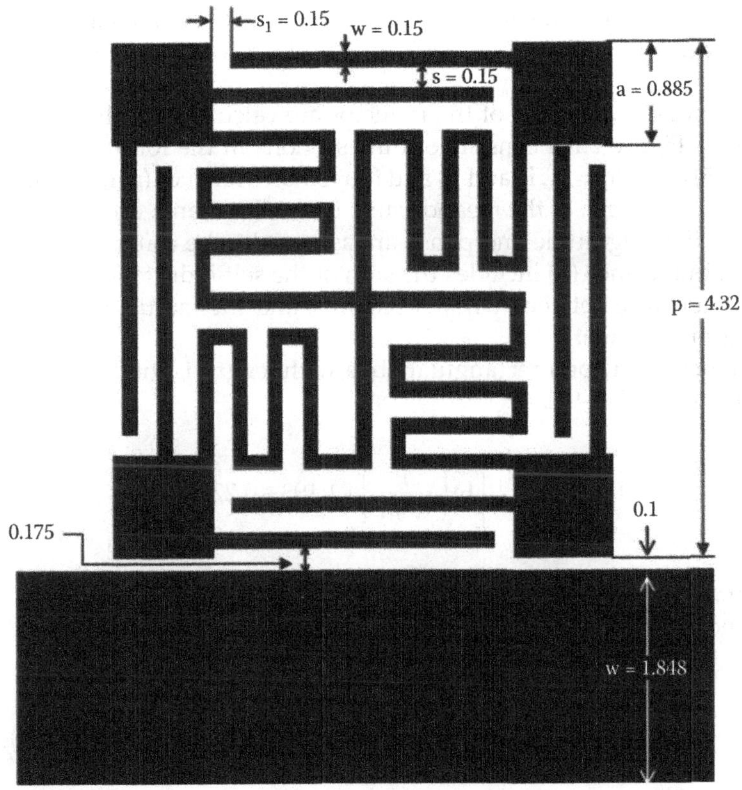

FIGURE 2.15
Single unit cell coupled to a microstrip transmission line.

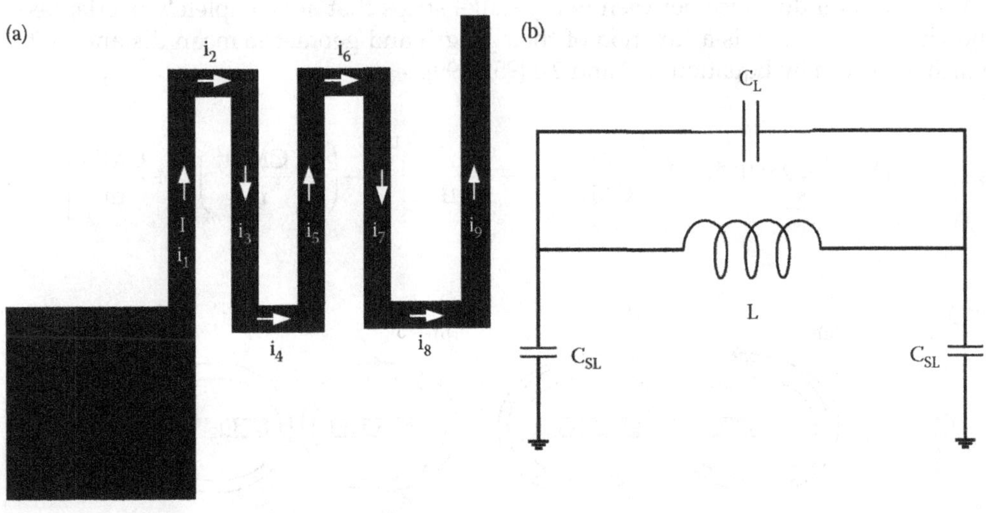

FIGURE 2.16
Meander line inductor. (a) Metal pattern. (b) Equivalent circuit model.

The pi-equivalent circuit of the meander line inductor is shown in Figure 2.16b. It consists of the inductor (L), parasitic capacitor (C_L) in parallel to L, and shunt capacitance (C_{SL}) to ground.

Equivalent circuit parameters of the inductor are calculated in the following way [95]. The inductor in Figure 2.16a consists of nine sections in the form of a rectangular strip with five (N = 5) arms (i_1, i_3, i_5, i_7, and i_9) and four small branches (i_2, i_4, i_6, and i_8) connecting the arms. The total length of the meander line is smaller than a wavelength (one-fifth the size), and thus the magnitude and phase are assumed to be constant along the meander line. The total inductance (L) includes the sum of the self-inductance of the nine sections, the mutual inductance between parallel sections, and the mutual inductance due to the effect of the ground plane.

The self-inductance L_i of a rectangular strip with length l, width w, and thickness t is given by Equation 2.1 [95–98].

$$L_i = 2\times10^{-4} l\left[\ln\left(\frac{l}{w+t}\right)+1.193+0.2235\frac{w+t}{l}\right] \tag{2.1}$$

where inductance L_i is in nH and all other dimensions are in μm.

Total self-inductance (L_s) is the sum of the self-inductance of each section:

$$L_s = \sum_{i=1}^{2N-1} L_i \tag{2.2}$$

The sections of the meander line that are parallel will have mutual inductance between them. Figure 2.17 shows the magnetic flux lines in positive and negative mutual inductance cases. If the current flowing through two parallel arms runs in the same direction, then the mutual inductance is positive, whereas if the current directions are opposite, then the mutual inductance is negative.

The mutual inductance between two parallel strips that are completely overlapped, as shown in Figure 2.18, is a function of their length and geometric mean distance (GMD), which are given by Equations 2.3 and 2.4 [95,96,99]:

$$M^{m,GMD} = 2\times10^{-4} m\left[\ln\left\{\frac{m}{GMD}+\left(1+\frac{m^2}{GMD^2}\right)^{1/2}\right\}-\left(1+\frac{GMD^2}{m^2}\right)^{1/2}+\frac{GMD}{m}\right] \tag{2.3}$$

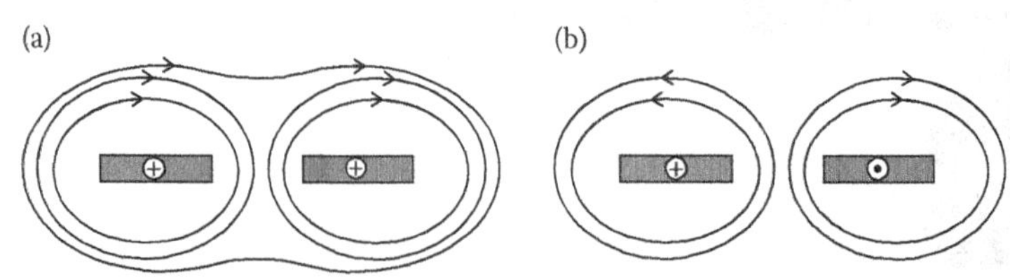

FIGURE 2.17
Magnetic flux lines. (a) Positive mutual inductance. (b) Negative mutual inductance.

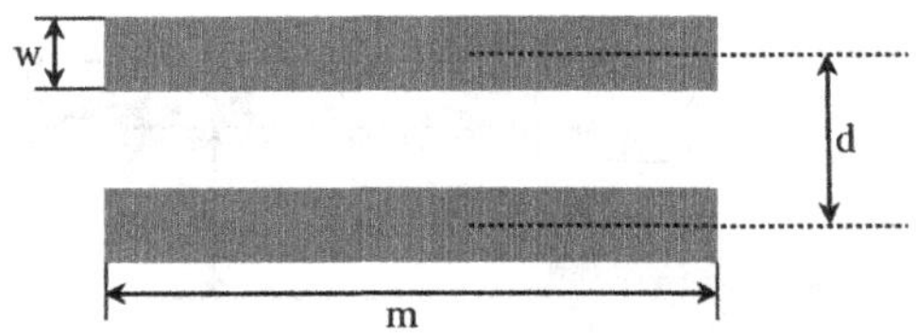

FIGURE 2.18
Parallel strips with complete overlap.

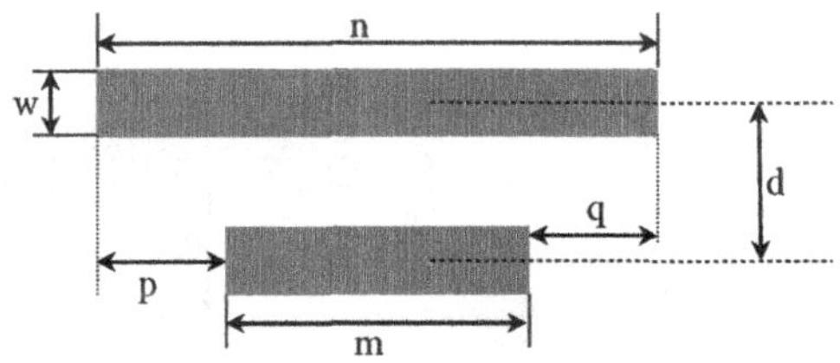

FIGURE 2.19
Parallel strips overlapped with unequal strip lengths.

$$\ln(\mathrm{GMD}) = \ln(d) - \left\{ \left[\frac{1}{12(d/w)^2}\right] + \left[\frac{1}{60(d/w)^4}\right] + \left[\frac{1}{168(d/w)^6}\right] + \left[\frac{1}{360(d/w)^8}\right] + \left[\frac{1}{660(d/w)^{10}}\right] + \cdots\cdots\cdots \right\} \tag{2.4}$$

where mutual inductance M is measured in nH, m (the length of the strip) in μm, and d (the center-to-center distance between the parallel strips) in μm.

If the parallel strips are overlapped with unequal strip lengths, as shown in Figure 2.19, the mutual inductance is given by Equation 2.5; if $p = 0$, then mutual inductance is given by Equation 2.6:

$$2M = M^{m+p,GMD} + M^{m+q,GMD} - M^{p,GMD} - M^{q,GMD} \tag{2.5}$$

$$2M = M^{m,GMD} + M^{n,GMD} - M^{q,GMD} \tag{2.6}$$

If the parallel strips are overlapped partially, as shown in Figure 2.20, the mutual inductance is given by Equation 2.7:

$$2M = M^{m+n-p,GMD} + M^{p,GMD} - M^{m-p,GMD} - M^{n-p,GMD} \tag{2.7}$$

If the parallel strips are not overlapped, as shown in Figure 2.21, the mutual inductance is given by Equation 2.8, and if p is zero, the mutual inductance is given by Equation 2.9:

$$2M = M^{m+n+p,GMD} + M^{p,GMD} - M^{m+p,GMD} - M^{n+p,GMD} \tag{2.8}$$

$$2M = M^{m+n,GMD} - M^{m,GMD} - M^{n,GMD} \tag{2.9}$$

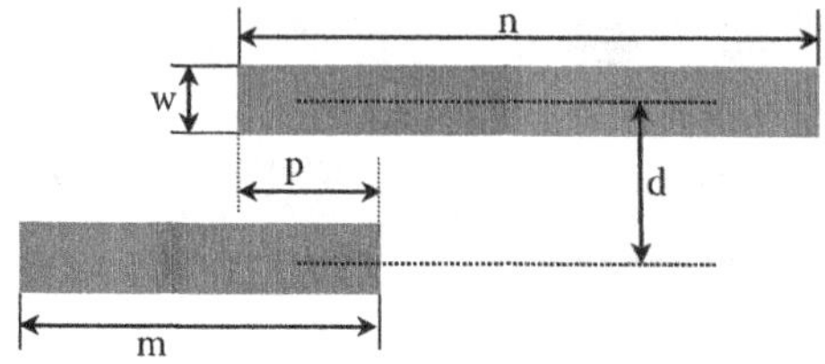

FIGURE 2.20
Parallel strips partially overlapped with unequal strip lengths.

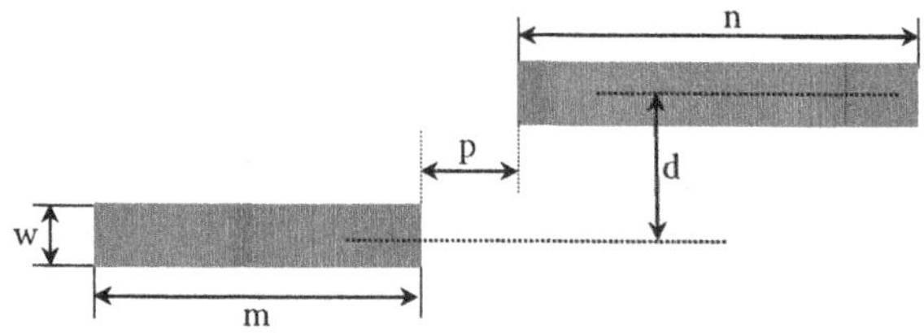

FIGURE 2.21
Parallel strips without overlap.

The mutual inductance between parallel arms of the inductor shown in Figure 2.16a is calculated based on their arrangement using Equations 2.3 through 2.9. Some of the mutual inductances are negative, such as $M_{1,3,}$ and some are positive, such as $M_{1,5}$. Because we assume that the magnitude and phase of the currents along the line are constant, $M_{1,3} = M_{3,1}$. Therefore, mutual inductance is calculated only once and then multiplied by two. Because the branches (i_2, i_4, i_6, and i_8) are not overlapped and the distance between the nonadjacent branches (i_2, i_6, and so on) is large, the mutual inductance is negligible and hence only calculated between adjacent branches ((i_2, i_4), (i_4, i_6), and (i_6, i_8)) by using Equation 2.9. The total mutual inductance is given by Equation 2.10:

$$L_M = 2\sum_{i=1}^{N-1}\sum_{j=1}^{N-i}(-1)^j M_{2i-1,2(i+j)-1} + 2\sum_{i}^{N-2} M_{2i,2i+2} \tag{2.10}$$

The effect of the ground plane is accounted for by using image theory. The image is located at twice the substrate thickness from the meander line. The mutual inductance between the meander line and its image is negative because the current flow in the image is opposite in the return path [100]. Since the distance between the inductor and its image is greater, GMD is approximated by the distance between the strips (twice the substrate thickness). Mutual inductance is calculated for each arm with other parallel arms in the image by using Equations 2.3 through 2.9. The total mutual inductance due to the image is given by Equation 2.11. The total inductance L is given by Equation 2.12 using Equations 2.2, 2.10, and 2.11:

$$L_{M\_image} = \sum_{i=1}^{2N-1}\left\{\begin{array}{ll}\displaystyle\sum_{j=1}^{N}(-1)^{\left(\frac{|i-(2j-1)|}{2}+1\right)} M_{i,2j-1} & \text{for i odd} \\ \displaystyle-\sum_{j=1}^{N-1} M_{i,2j} & \text{for i even}\end{array}\right\} \tag{2.11}$$

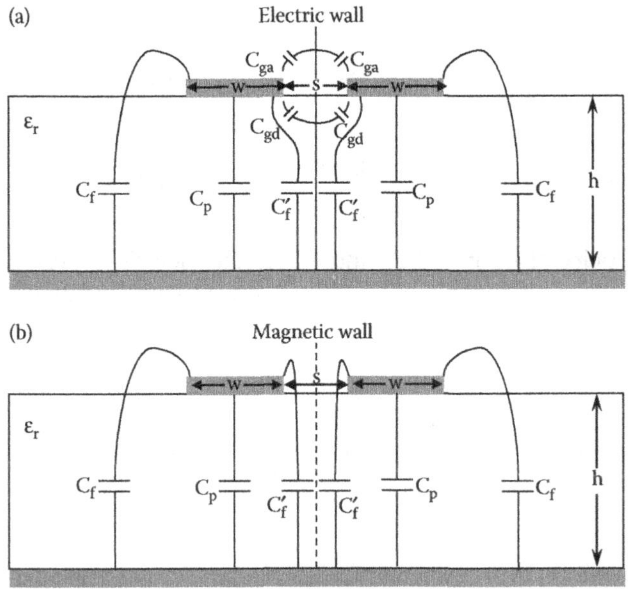

FIGURE 2.22
Symmetric coupled microstrip lines. (a) Odd-mode capacitance. (b) Even-mode capacitance.

$$L = L_s + L_M + L_{M\_image} \tag{2.12}$$

The parasitic capacitance C_L in the equivalent circuit of the inductor in Figure 2.16b is the equivalent capacitance between the adjacent arms of the inductor. Symmetric coupled line theory is used to calculate the capacitance between two adjacent arms of the inductor [101]. The even-mode and odd-mode capacitance of the symmetric coupled microstrip lines of width w and separation s are shown in Figure 2.22.

As shown in Figure 2.22, the odd-mode capacitance per unit length (C_o) is given by Equation 2.13, and even-mode capacitance per unit length is given by Equation 2.14:

$$C_o = C_p + C_f + C'_f + C_{gd} + C_{ga} \tag{2.13}$$

$$C_e = C_p + C_f + C'_f \tag{2.14}$$

C_p is the parallel plate capacitance per unit length between the line and ground plane (as calculated by Equation 2.15), C_f is the fringe capacitance per unit length of a microstrip line to the ground at the outer edge of the microstrip line (given by Equation 2.16), and C'_f (given in Equations 2.17 and 2.18) is the modified fringe capacitance of a microstrip line to the ground in the presence of another line separated by distance s:

$$C_p = \varepsilon_0 \varepsilon_r \frac{w}{h} \tag{2.15}$$

$$C_f = 0.5\left(\frac{\sqrt{\varepsilon_{re}}}{(cZ_o) - C_p}\right) \tag{2.16}$$

$$C_f' = \frac{C_f}{1 + A(h/s)\tanh(10\,s/h)}\left(\frac{\varepsilon_r}{\varepsilon_{re}}\right)^{1/4} \tag{2.17}$$

$$A = \exp\left(-0.1\exp\left(2.33 - 1.5\frac{w}{h}\right)\right) \tag{2.18}$$

where c is the velocity of light and Z_0 and ε_{re} are characteristic impedance and effective permittivity of the microstrip line with width w and substrate height h, respectively (given by Equations 2.19 through 2.22).

For $w/h \le 1$:

$$\varepsilon_{re} = \frac{\varepsilon_r + 1}{2} + \frac{\varepsilon_r - 1}{2}\left\{\left(1 + 12\frac{h}{w}\right)^{-0.5} + 0.04\left(1 - \frac{w}{h}\right)^2\right\} \tag{2.19}$$

For $w/h \ge 1$:

$$\varepsilon_{re} = \frac{\varepsilon_r + 1}{2} + \frac{\varepsilon_r - 1}{2}\left\{\left(1 + 12\frac{h}{w}\right)^{-0.5}\right\} \tag{2.20}$$

For $w/h \le 1$:

$$Z_o = \frac{120\pi}{2\pi\sqrt{\varepsilon_{re}}}\ln\left(\frac{8h}{w} + 0.25\frac{w}{h}\right) \tag{2.21}$$

For $w/h \ge 1$:

$$Z_o = \frac{120\pi}{\sqrt{\varepsilon_{re}}}\left\{\frac{w}{h} + 1.393 + 0.677\ln\left(\frac{w}{h} + 1.444\right)\right\}^{-1} \tag{2.22}$$

C_{gd} is the fringing capacitance per unit length between the symmetric coupled microstrip lines through a dielectric, and it is calculated from the corresponding coupled stripline geometry with double the substrate thickness spacing between the ground planes (given in Equation 2.23):

$$C_{gd} = 0.5(C_{os} - C_{es}) \tag{2.23}$$

C_{os} and C_{es} are the odd- and even-mode capacitances per unit length of the symmetric coupled stripline, given by Equations 2.24 and 2.25:

$$C_{os} = 4\varepsilon_0\varepsilon_r\frac{K(k_o)}{K(k_o')} \tag{2.24}$$

$$C_{es} = 4\varepsilon_0\varepsilon_r\frac{K(k_e)}{K(k_e')} \tag{2.25}$$

K(x) is the elliptic function, and k_o and k_e are given by Equations 2.26 and 2.27:

$$k_o = \tanh\left(\frac{\pi}{4}\frac{w}{h}\right)\coth\left(\frac{\pi}{4}\frac{w+s}{h}\right) \quad \text{and} \quad k_o' = \sqrt{1-k_o^2} \tag{2.26}$$

$$k_e = \tanh\left(\frac{\pi}{4}\frac{w}{h}\right)\tanh\left(\frac{\pi}{4}\frac{w+s}{h}\right) \quad \text{and} \quad k_e' = \sqrt{1-k_e^2} \tag{2.27}$$

C_{ga}, the fringing capacitance per unit length between the symmetric coupled microstrip lines, is calculated from the capacitance of the corresponding coplanar strips with air as a dielectric [101,102] and is given in Equations 2.28 and 2.29:

$$C_{ga} = 0.5C_{cps}^{a} - \Delta C_f \tag{2.28}$$

$$\Delta C_f = (C_f^a - C_{fs}^a) + (C_f'^a - C_{fs}'^a) = (C_f^a + C_f'^a) - 0.5C_{es}^a + C_{ps}^a \tag{2.29}$$

C_{cps}^a is the total fringing capacitance per unit length between the strips of a symmetric coupled coplanar strip (CPS) with air as a dielectric, as given by Equation 2.30:

$$C_{cps}^a = 2\varepsilon_0\frac{K(k')}{K(k)} \quad \text{with} \quad k = \frac{s}{(s+2w)} \quad \text{and} \quad k' = \sqrt{1-k^2} \tag{2.30}$$

$C_f^a, C_f'^a$ are the fringing capacitances per unit length of a symmetric coupled microstrip line with air as a dielectric, calculated using Equations 2.16 through 2.18 with $\varepsilon_r = 1$.

C_{es}^a is the even-mode capacitance per unit length of a symmetric coupled stripline with air as a dielectric, calculated using Equation 2.25 with $\varepsilon_r = 1$.

$C_{ps}^a = C_p^a$ is the parallel plate capacitances per unit length of a symmetric coupled stripline with air as a dielectric, calculated using Equation 2.15 with $\varepsilon_r = 1$.

The total capacitance between the two coupled lines in Figure 2.22 is given by Equation 2.31:

$$C_g = 0.5 * l\,(C_{ga} + C_{gd}) \tag{2.31}$$

where l is the effective length (minimum length of the two lines).

The parasitic capacitance (C_L) of the inductor is the equivalent of series interarm capacitance of the meander inductor and is given by Equation 2.32:

$$C_L = \frac{1}{\sum_{i=1}^{N-1}(1/C_{gi})} \tag{2.32}$$

where C_{gi} is the ith interarm capacitance of the inductor, given by Equation 2.31.

The shunt parasitic capacitance (C_{SL}) to the ground in the equivalent circuit of the inductor in Figure 2.16b is calculated from the parallel plate and fringing capacitance of each line of the inductor. The branches in the inductor are uncoupled microstrip lines, and therefore, the total shunt capacitance to the ground per unit length is $C_s = C_p + 2C_f$. The shunt capacitance per unit length of the coupled microstrip line is $C_s = C_p + C_f + C_f'$. In the case of three symmetric coupled microstrip lines, the center line is coupled on both sides and the shunt capacitance to the ground per unit length is $C_s = C_p + 2C_f'$. The total shunt capacitance is represented by two shunt capacitors (C_{SL}) in the equivalent circuit (Figure 2.16b). C_{SL} is given by Equation 2.33:

$$C_{SL} = 0.5\sum_{i=1}^{2N-1} l_i C_{si} \tag{2.33}$$

where C_{Si} is the shunt capacitance to the ground of the ith section of the inductor.

The small pad of width Pw to which the meander line is connected is wider than the meander line itself. Therefore, the pad has low impedance whose equivalent circuit is a shunt capacitor. The length of the pad is small, so the shunt capacitance is approximated using Equations 2.34 and 2.35:

$$B \approx \frac{2\pi Pw}{\lambda_g Z_p} \tag{2.34}$$

$$C_{ph} \approx Pw * \frac{\sqrt{\varepsilon_{rep}}}{(Z_{po} * c)} \tag{2.35}$$

where c is the velocity of light and Z_{po} and ε_{rep} are the characteristic impedance and effective permittivity, calculated using Equations 2.19 through 2.22 with pad width Pw and substrate height h as microstrip parameters.

The metal pattern of the interdigital capacitor of the unit cell and its equivalent circuit are shown in Figure 2.23. The expression for the capacitance of an interdigital capacitor is given by Equation 2.36 [103]:

$$C = \frac{\varepsilon_{re} 10^{-3}}{18\pi} \frac{K(k)}{K'(k)} (N-1)l \quad (pF) \tag{2.36}$$

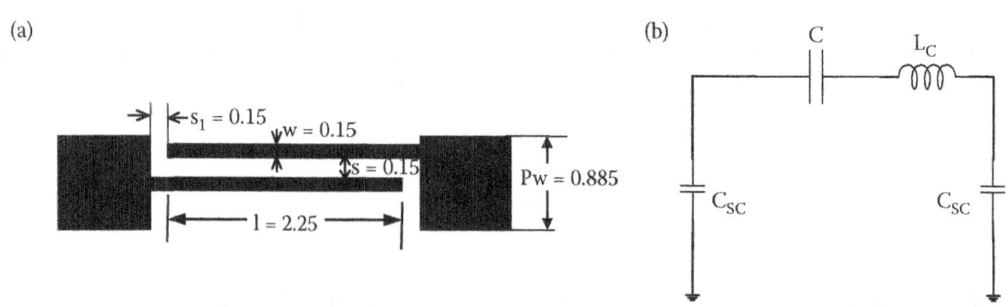

FIGURE 2.23
Interdigital capacitor. (a) Metal pattern. (b) Equivalent circuit.

where l is the length in μm, N is the number of fingers, and ε_{re} is the effective relative permittivity of the microstrip line of width w. The ratio of the elliptic integral of the first kind K(k) and its complement K′(k) is given by Equations 2.37 and 2.38:

$$\frac{K(k)}{K'(k)} = \begin{cases} \frac{1}{\pi}\ln\left(2\frac{1+\sqrt{k}}{1-\sqrt{k}}\right) & \text{for } 0.707 \le k \le 1 \\ \frac{\pi}{\ln\left(2\frac{\left(1+\sqrt{k'}\right)}{\left(1-\sqrt{k'}\right)}\right)} & \text{for } 0 \le k \le 0.707 \end{cases} \tag{2.37}$$

where

$$k = \tan^2\left(\frac{a\pi}{4b}\right), \quad a = w/2, \quad b = (w+s)/2, \quad \text{and} \quad k' = \sqrt{1-k^2} \tag{2.38}$$

The parasitic inductance (L_C) and shunt capacitance (C_{SC}) are given by Equations 2.39 and 2.40 [103]:

$$L_C = \frac{Z_0\sqrt{\varepsilon_{re}}}{c} l \tag{2.39}$$

$$C_{SC} = \frac{1}{2}\frac{\sqrt{\varepsilon_{re}}}{Z_0 c} l \tag{2.40}$$

where c is the velocity of light in free space and Z_0 and ε_{re} are calculated using Equations 2.19 through 2.22 with w′ = (w + s)N − s and h as microstrip parameters.

The calculation of coupling capacitance between two asymmetric microstrip lines (pad of width Pw and transmission line of width Lw), separated by spacing s and shown in Figure 2.24, is as follows [95,104].

The total capacitance between the two microstrip lines (pad and transmission line) is given by Equation 2.41:

$$C_{m\_PL} = \left(C_{ga\_PL} + C_{gd\_PL}\right) * l \tag{2.41}$$

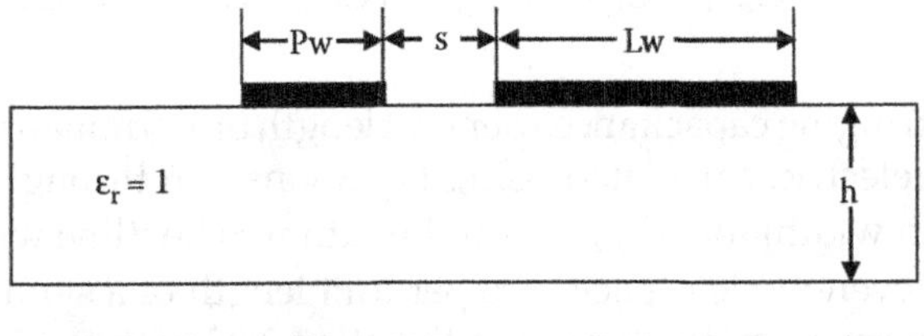

FIGURE 2.24
Asymmetric coupled microstrip transmission line.

where l is the overlap length of the two lines. C_{gd_PL} is the total fringing capacitance per unit length between the two symmetric coupled lines (pad and line) through a dielectric and is calculated from Equation 2.42:

$$C_{gd\_PL} = 0.5\sqrt{C_{gd\_P}C_{gd\_L}} \tag{2.42}$$

where $0.5C_{gd\_P}$ is the total fringing capacitance per unit length between the two symmetric coupled microstrip lines of width Pw through a dielectric. Similarly, $0.5C_{gd\_L}$ is the total fringing capacitance per unit length between the two symmetric coupled microstrip lines of width Lw through a dielectric. C_{gd_P} and C_{gd_L} can be calculated using Equations 2.23 through 2.27 with widths w = Pw and Lw, respectively.

C_{ga_PL}, the total fringing capacitance per unit length between the two asymmetric coupled microstrip lines (pad and line) through air, is given by Equation 2.43:

$$C_{ga\_PL} = 0.5C^{a}_{CPS\_PL} - \Delta C_{f\_PL} \tag{2.43}$$

where $C^{a}_{CPS\_PL}$ is the total fringing capacitance per unit length between asymmetric coplanar strips of pad width (Pw) and line width (Lw), separated by s and suspended in air. This calculation is given by Equations 2.44 and 2.45 [101]:

$$C^{a}_{CPS\_PL} = 2\varepsilon_0 K(k')/K(k) \tag{2.44}$$

where,

$$k = \sqrt{\frac{1+\frac{Pw}{s}+\frac{Lw}{s}}{\left(1+\frac{Pw}{s}\right)\left(1+\frac{Lw}{s}\right)}} \quad \text{and} \quad k = \sqrt{1-k^2} \tag{2.45}$$

ΔC_{f_PL} is the total fringing capacitance in the surrounding air, as given by Equations 2.46 through 2.48:

$$\Delta C_{f\_PL} = \frac{\Delta C_{f\_P} * \Delta C_{fL}}{\Delta C_{f\_P} + \Delta C_{fL}} \tag{2.46}$$

$$\Delta C_{f\_P} = \left(C^{a}_{f\_P} - C^{a}_{fs\_P} + C'^{a}_{f\_P} - C'^{a}_{fs\_P}\right) = \left(C^{a}_{f\_P} + C'^{a}_{f\_P}\right) - 0.5C^{a}_{es\_P} + C^{a}_{ps\_P} \tag{2.47}$$

$$\Delta C_{f\_L} = \left(C^{a}_{f\_L} - C^{a}_{fs\_L} + C'^{a}_{f\_L} - C'^{a}_{fs\_L}\right) = \left(C^{a}_{f\_L} + C'^{a}_{f\_L}\right) - 0.5C^{a}_{es\_L} + C^{a}_{ps\_L} \tag{2.48}$$

$C^{a}_{f\_P}$, $C'^{a}_{f\_P}$ are the fringing capacitances per unit length of a symmetric coupled microstrip line with air as a dielectric, calculated using Equations 2.16 through 2.18 with $\varepsilon_r = 1$ and width w = Pw (patch width) for $C^{a}_{f\_P}$, $C'^{a}_{f\_P}$ and width w = Lw (line width) for $C^{a}_{f\_L}$, $C'^{a}_{f\_L}$.

$C^{a}_{es\_P}$, $C^{a}_{es\_L}$ is the even-mode capacitance per unit length of a symmetric coupled stripline with air as a dielectric, calculated using Equation 2.25 with $\varepsilon_r = 3.2$ and width w = Pw (patch width) for $C^{a}_{es\_P}$ and width w = Lw (line width) for $C^{a}_{es\_L}$.

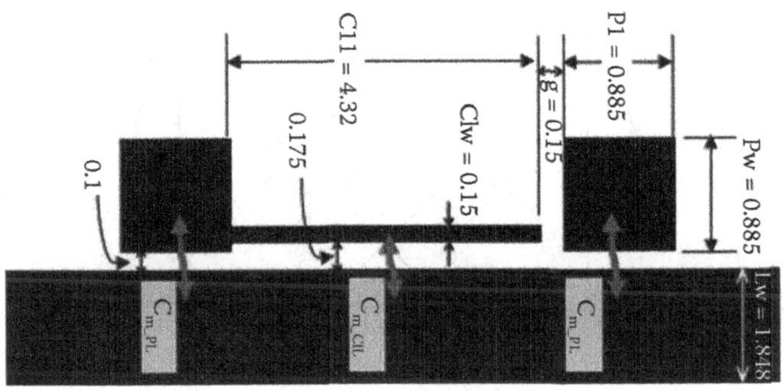

FIGURE 2.25
Coupling capacitance between transmission line and the unit cell.

$C^{a}_{ps\_P} = C^{a}_{p\_P}$, $C^{a}_{ps\_L} = C^{a}_{p\_L}$ are the fringing capacitances per unit length of a symmetric coupled stripline with air as dielectric, calculated using Equation 2.15 with $\varepsilon_r = 1$ and width w = Pw for $C^{a}_{p\_P}$ and width w = Lw for $C^{a}_{p\_L}$.

Figure 2.25 shows various components of coupling capacitance between the unit cell and transmission line. The coupling capacitance is calculated by separately applying asymmetric coupled line theory to the pads of the unit cell and the finger of the capacitor coupled to the transmission line. C_{m_PL} is the coupling capacitance between the asymmetric coupled transmission line of width 1.848 mm and the pad of width 0.885 mm, separated by 0.1 mm. C_{m_CIL} is the coupling capacitance between the asymmetric coupled transmission line of width 1.848 mm and the capacitor line of width 0.15 mm, separated by 0.175 mm. C_{m_PL} and C_{m_CIL} are calculated using Equation 2.41 through 2.48 with their corresponding parameters. The total coupling capacitance C_m is given by Equation 2.49:

$$C_m = 2C_{m\_PL} + C_{m\_CIL} \tag{2.49}$$

The circuit parameters of the inductor are L, C_L, and C_{SL}. The circuit parameters of the capacitor are C, L_C, and C_{SC}. The parasitic shunt capacitance of each pad to the ground is C_{ph}. The coupling capacitance between the cell and the transmission line is C_m. Arranging all the parameters according to the geometry of the unit cell yields the final circuit, as shown in Figure 2.26. Here, the shunt capacitance C_{S1} to the ground at the center of the circuit is $4C_{SL}$. The shunt capacitance to the ground at each corner (C_S) is $C_{SL} + 2C_{SC} + C_{ph}$. The coupling capacitance between the unit cell and transmission line is represented by placing two connected capacitors (C_C) between the line and the unit cell in the circuit and is given by $C_c = C_m/2$.

By using the above procedure, the equivalent circuit parameters of the unit cell coupled to the transmission line shown in Figure 2.15 are calculated. The values are L = 2.5156 nH, C_L = 4.56 fF, C = 45.1 fF, L_C = 1.177 nH, C_S = 0.2328 pF, C_{S1} = 0.240 pF, and C_C = 54.5 fF. A short microstrip transmission line of length 4.32 mm is placed between the two C_Cs to account for phase. The comparison of electromagnetic (EM) simulation and circuit simulation is shown in Figure 2.27. The resonance frequencies in the EM simulation and the circuit simulation are 5.34 GHz and 5.3 GHz, respectively. A 40 MHz difference is observed between the circuit- and EM-simulated results. The equivalent circuit approach is helpful in choosing dimensions of the unit cell, and further optimization can be achieved using EM simulators.

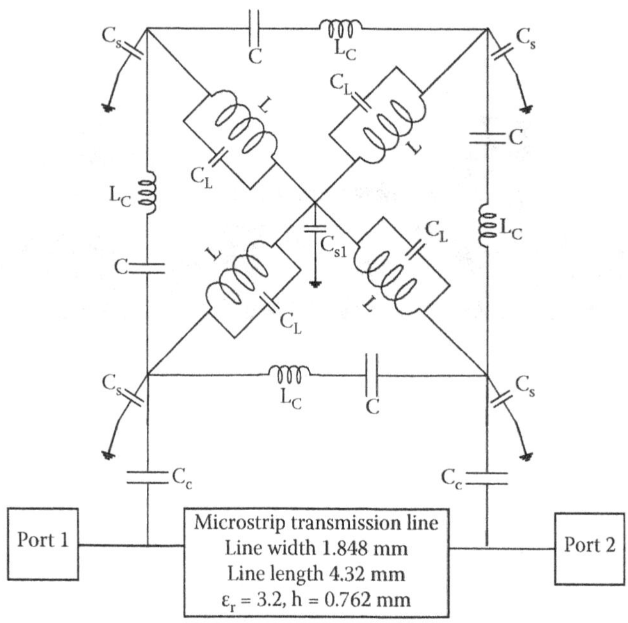

FIGURE 2.26
Equivalent circuit of a single unit cell coupled to a transmission line.

The equivalent circuit presented in this book can predict only the bandgap position, not the entire set of characteristics of the stopband, such as bandwidth and depth. Thus, the resistive elements are not considered in the equivalent circuit.

As observed, the coupling of a single cell to a transmission line results in a notch with a low rejection level. By coupling two one-dimensional arrays (1×7) of these cells on either side of the transmission line, as described in Section 2.5, the rejection level of the

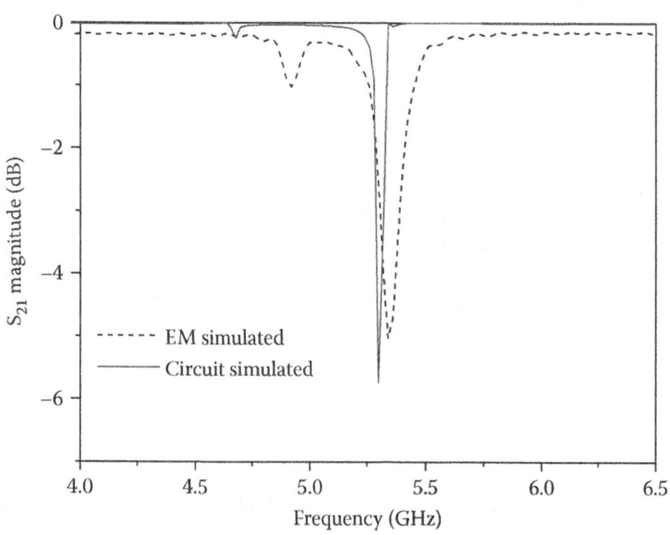

FIGURE 2.27
Comparison of circuit- and EM-simulated results of a single cell coupled to a transmission line.

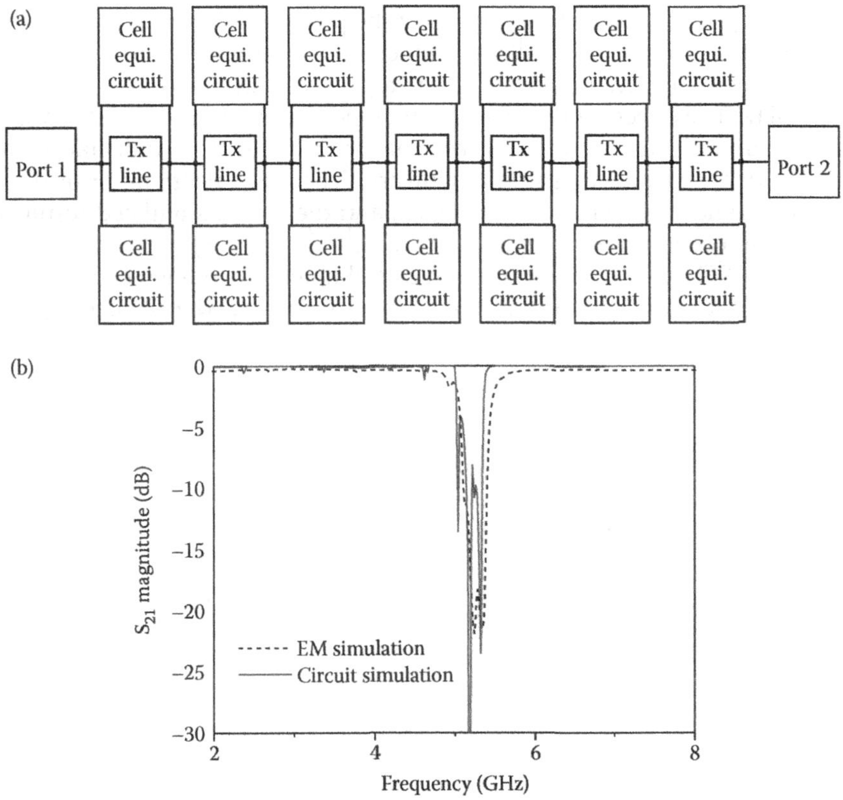

FIGURE 2.28
Band-stop filter. (a) Equivalent circuit. (b) Comparison of circuit- and EM-simulated results.

band-stop filter is improved, as shown in Figure 2.9c. The equivalent circuit of the band-stop filter shown in Figure 2.28a is derived using the equivalent circuit of the single cell. Each "Cell Equi. Circuit" block represents the equivalent circuit of the single cell shown in Figure 2.26, and each equivalent circuit of the single cell is connected across the small microstrip transmission line of length 4.32 mm. The circuit simulation is done in ADS software, and its comparison with EM simulation is shown in Figure 2.28b. It is observed that the circuit-simulated results match well with the EM-simulated results.

2.7 Conclusion

In this chapter, a UC-EBG structure was developed with meander lines and interdigital capacitors. The structure was characterized by a dispersion diagram using the Eigenmode solver of CST Microwave Studio. In the dispersion diagram, the bandgap of the proposed structure existed from 5.48 to 7.9 GHz, whereas the bandgap of a conventional planar EBG structure with the same size and substrate parameters ranged from 9.41 to 12.41 GHz. A 38% reduction in the bandgap center frequency was seen in the proposed structure compared to the conventional UC-EBG. Surface wave measurements performed on an array of 7 × 7 unit cells showed the bandgap of the proposed structure from 5.74 to 8 GHz. A one-dimensional array of EBG cells was edge-coupled to a microstrip line on either side of the transmission line, resulting in a band-stop/notch filter with 3 dB bandwidth from 4.95 to 5.37 GHz. The effects of the meander line inductor and interdigital capacitor of the unit cell were investigated, and it was seen that with increases in inductance and capacitance, the resonant frequency shifts downward. The equivalent circuit of the unit cell coupled to the transmission line was modeled, and the circuit-simulated results matched the EM-simulated results well.

By using the EBG notch filter reported in this chapter, a band-notched UWB filter and a band-notched ultra wide band (UWB) antenna are developed, tested, and presented in Chapter 3.

3

Compact Planar Periodic Structure as Band-Rejection Filter

3.1 Introduction

The bandstop/notch filter reported in Chapter 2 using EBG structures can be used to create a band-notched ultra wideband (UWB) filter by cascading it with a UWB filter.

A UWB system operates from 3.1 to 10.6 GHz. In such a wide band, the UWB systems can get interference from existing narrow band systems such as WLAN systems working in 5.15 to 5.35 GHz, 5.725 to 5.825 GHz bands, WiMax operating at 3.5 GHz, and satellite communication bands operating around 8 GHz. Thus, it is desirable to have a notch at those bands in the UWB band-pass filter (BPF) to avoid interference.

UWB BPFs with notched bands have been reported using mushroom-type electromagnetic bandgap structures [73], using a simplified composite right/left-handed (SCRLH) resonator [74], embedding a stepped impedance resonator (SIR) [75,105], loading a multimode resonator (MMR) with a parasitic coupled line [106], using a split-ring resonator-defected ground structure (DSRR) [107], and using complementary split-ring resonators (CSRRs) [108].

To avoid interference, UWB antennas are designed with a notch using mushroom EBG structures, as proposed earlier in References 69–72, and recently a band notched UWB antenna using a planar interdigital capacitance loading loop resonator is proposed in Reference 77.

In Chapter 2, it is seen that by loading EBG cells on a transmission line, a narrow band-notch bandgap can be obtained. This is used to create a notch in the UWB filter.

3.2 MMR UWB Filter

The multiple-mode resonator (MMR) [109] consists of a low impedance line in the middle and two high impedance lines at the ends. The length of the low impedance line is half guide wavelength ($\lambda_g/2$) long and the high impedance lines are quarter guide wavelength ($\lambda_g/4$) long at the center frequency of the UWB band. By weakly coupling the MMR to two 50-ohm (Ω) lines as shown in Figure 3.1, resonant frequencies of the MMR can be known. Simulations are done using CST Microwave Studio software with a GML 1000 substrate ($\varepsilon_r = 3.2$, $\tan\delta = 0.002$, and thickness $h = 0.762$ mm). The simulated transmission coefficient showing resonant frequencies of the MMR is plotted in Figure 3.2. The first three resonant frequencies occur at 4.4 GHz, 6.84 GHz, and 9.33 GHz.

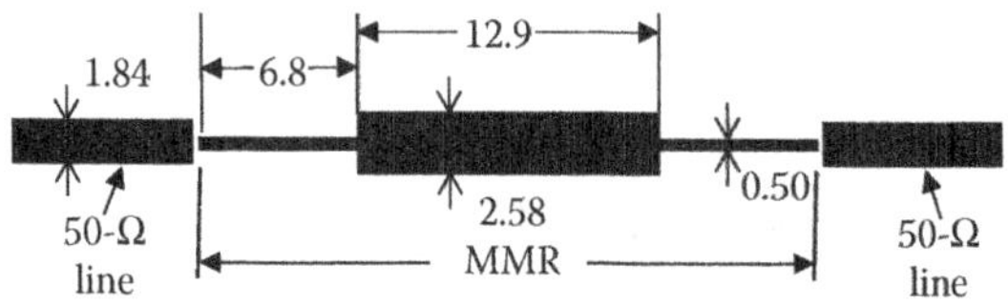

FIGURE 3.1
MMR weakly coupled to 50-Ω transmission lines. (All dimensions are in mm.)

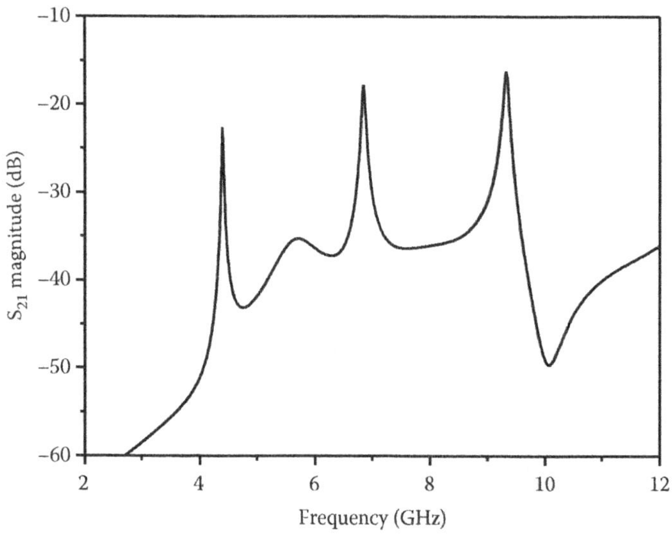

FIGURE 3.2
Resonant frequencies of MMR.

By strongly coupling MMR to the 50-Ω transmission line, $|S_{21}|$ can be made close to 0 dB around the three resonance frequencies of the MMR, resulting in an ultra-wide passband. For strong coupling, two parallel lines 0.2-mm wide connected to the 50-Ω lines are capacitively coupled to the high impedance lines of the MMR [110,111] as shown in Figure 3.3a. The fabricated filter photograph is shown in Figure 3.3b. A ZVB20 Rohde & Schwarz vector network analyzer is used for S-parameter measurements. Figure 3.3c shows the simulated and measured results. The measured 3-dB passband of the filter is from 3.47 GHz to 10.23 GHz. At 6.825 GHz, the measured insertion loss is 0.67 dB whereas the simulated insertion loss is 0.31 dB. The increased loss observed in the measured results can be attributed to connector loss. The return loss is better than 11.5 dB in the passband with a maximum group delay variation of 0.36 ns.

3.3 Band-Notched UWB Filter

Single band-notched and dual band-notched UWB filters are developed by cascading the MMR UWB filter of the previous section and the EBG bandstop/notch filter from the previous chapter. Three variations of single band-notched UWB filters (filter "A," "B," and "C")

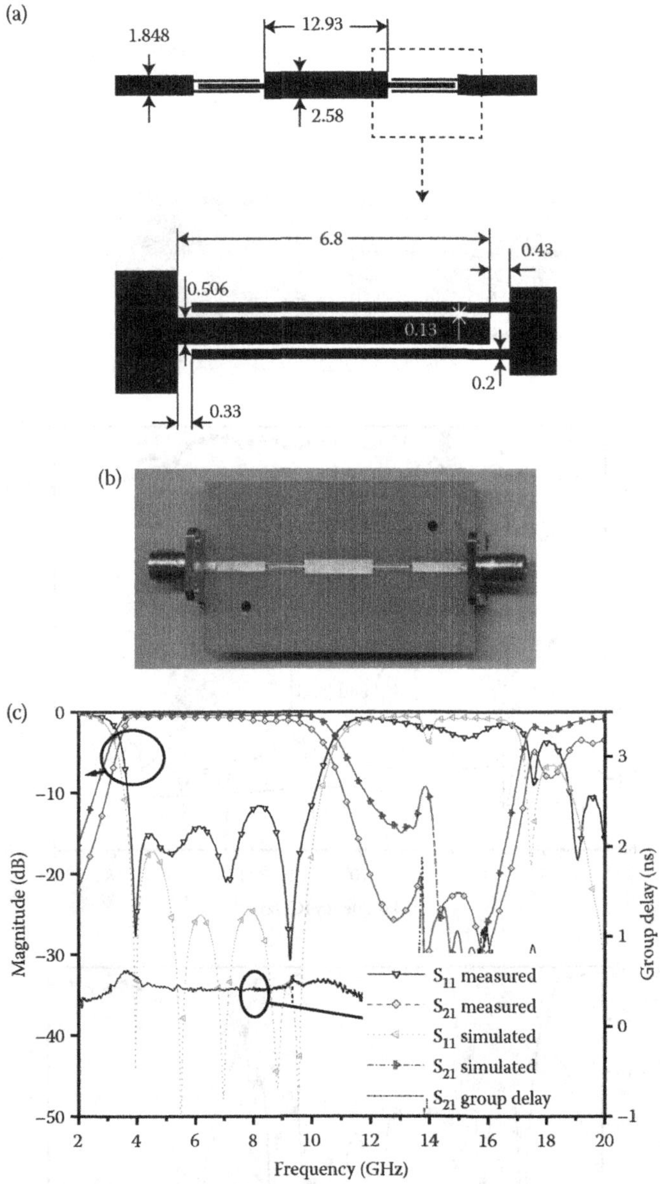

FIGURE 3.3
MMR UWB filter. (a) Schematic diagram, (b) photograph, and (c) measured and simulated results.

are discussed in Section 3.3.1 and a dual band-notched UWB filter (filter "D") is discussed in 3.3.2.

3.3.1 Single Band-Notched UWB Filter

In filter "A," seven unit cells are loaded on either side of the output line of the MMR filter as seen in Figure 3.4a. The proposed unit cell (Figure 2.1) described in Chapter 2 is used in this section. The effective length (from the edge of the MMR to the edge of notch circuit) of

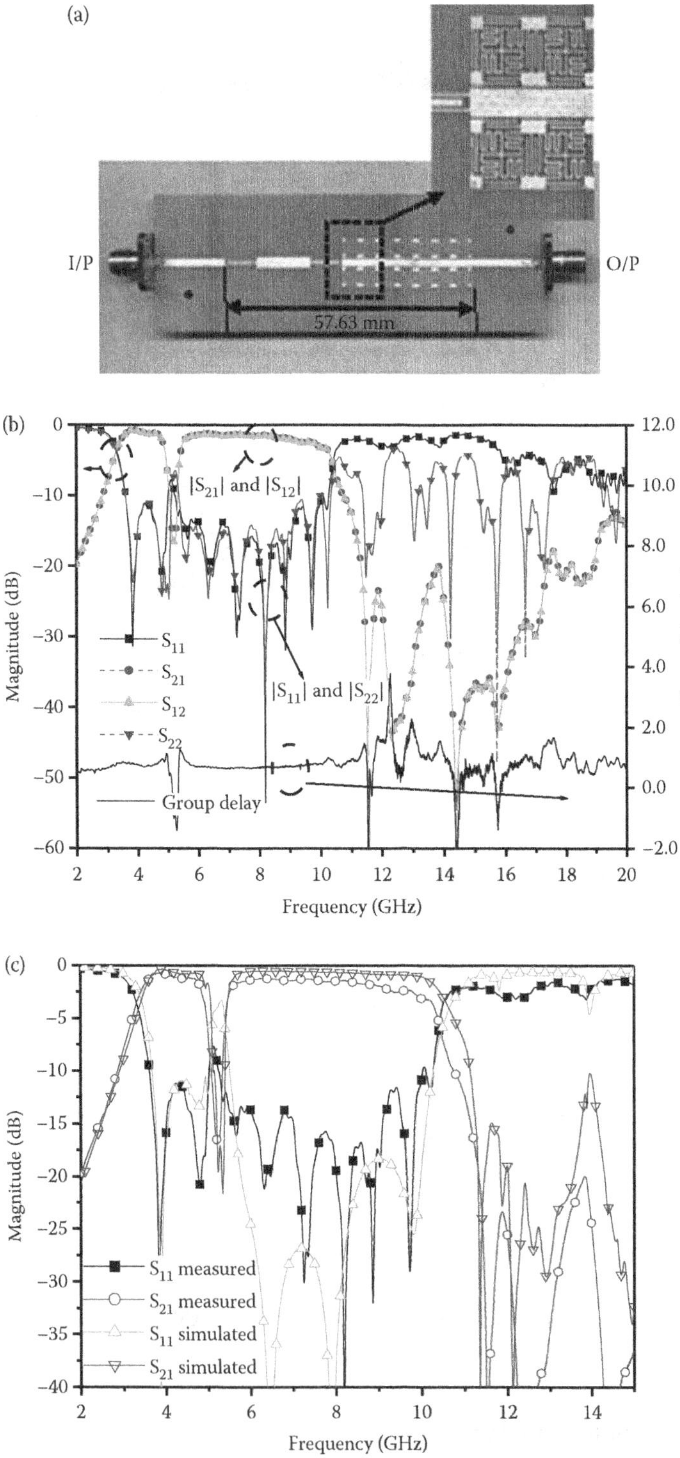

FIGURE 3.4
Band-notched UWB filter with EBG coupled to output line (Filter A): (a) photograph, (b) measured results, and (c) comparison of measured and simulated results.

filter "A" is 57.63 mm ($1.592\lambda_g$, λ_g at 5.19 GHz) as shown in Figure 3.4a. From the measured results shown in Figure 3.4b, the 3-dB passband is from 3.37 GHz to 9.97 GHz with a fractional bandwidth of 99%. A 3-dB notch band is realized from 4.88 to 5.48 GHz centered at 5.19 GHz with a rejection depth of 16.8 dB. The return loss is better than 11 dB in the passband except in the notch region. The maximum group delay variation in the passband is 0.28 ns before the notch and 0.54 ns after the notch. Comparison of the measured and simulated results is shown in Figure 3.4c.

The measured notch band is slightly shifted down when compared to the simulated results. This is because of the over-etching during fabrication of the filter. Due to over-etching, the meander lines become narrow which results in high inductance and a downward shift in the notch frequency. The measured insertion loss at the center frequency (6.67 GHz) is 1.35 dB, whereas the simulated insertion loss is 0.53 dB. The difference is mainly due to the connector losses that are not taken into account in the simulations. The transmission above 17 GHz in the band-notched UWB filter is below 10 dB, extending the upper stopband when compared to the UWB filter response in Figure 3.3c. This is because the higher bandgaps of the EBG structure fall in the upper stopband of the UWB filter. It is also seen from the band-notched UWB filter results plotted in Figure 3.4b that $|S_{11}|$ is close to 0 dB, but $|S_{22}|$ is below −10 dB at some points above 10 GHz, even though transmission from port 2 to port 1 ($|S_{12}|$) is well below −20 dB. This is because the EBG structure is radiating at higher frequencies. The same is not observed in the $|S_{11}|$ plot as the filter structure is preceding the EBG structure with input at port 1.

To study the interference of these radiations, simulations were done in CST Microwave Studio with open add space boundary conditions and electric ($E_t = 0$) boundary conditions in all directions. The simulated results shown in Figure 3.5 demonstrate that there is no difference in the performance of the filter, which shows that radiation due to the EBG structures at higher frequencies is not affecting the filter performance.

In filter "A," the notch depth is 16.8 dB. To improve the notch depth further, the EBG structure is coupled to both input and output lines of the MMR filter as shown in Figure 3.6a

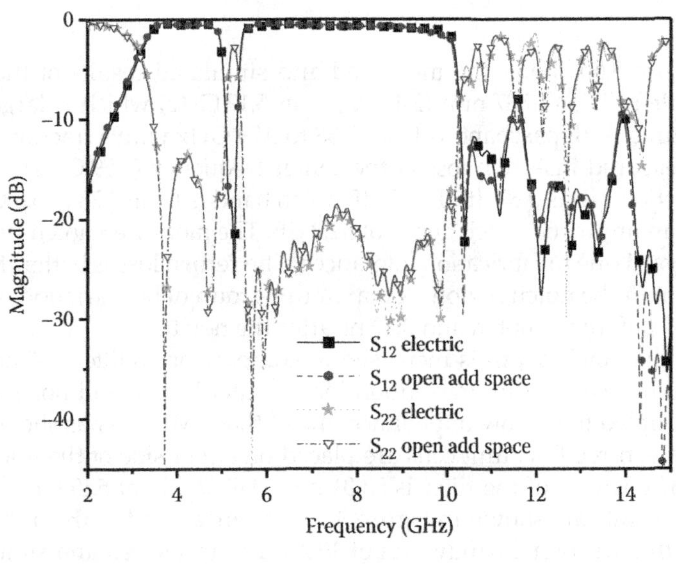

FIGURE 3.5
Simulated results of band-notched UWB filter with electric and open add space boundary conditions.

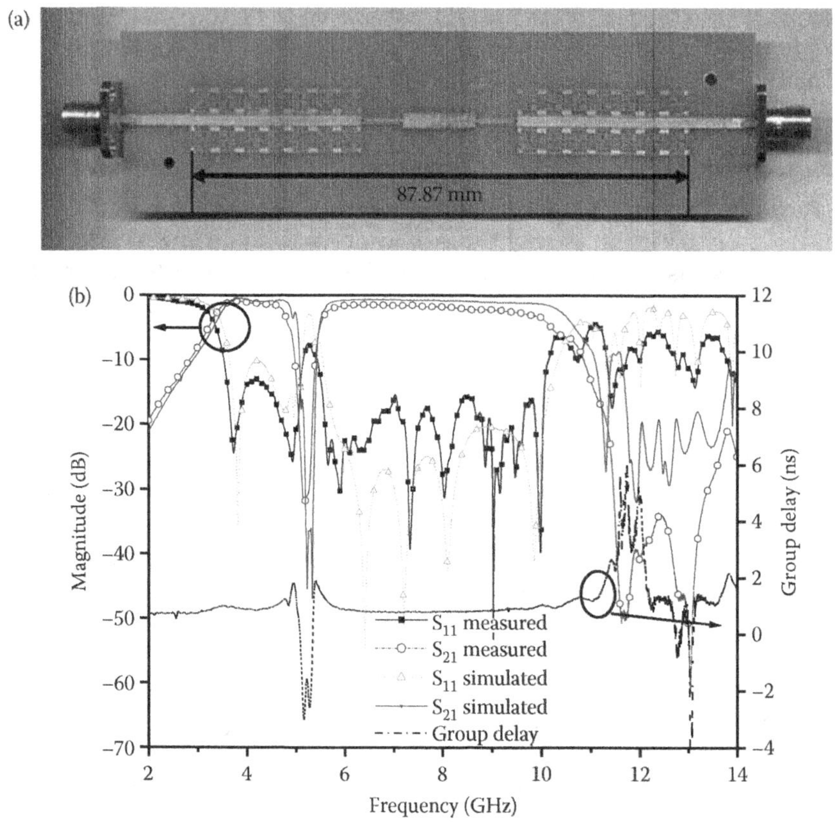

FIGURE 3.6
Band-notched UWB filter with EBG structure coupled to input and output lines (Filter B): (a) photograph and (b) measured and simulated results.

(filter "B"). Figure 3.6b shows the measured and simulated results of the filter "B." The length of the filter "B" is 87.87 mm ($2.427\lambda_g$, λ_g at 5.19 GHz) which is larger than in filter "A." The measured 3-dB passband is from 3.38 to 9.79 GHz with a fractional bandwidth of 97.27%. The measured insertion loss at the center frequency (6.59 GHz) is 1.5 dB and the simulated insertion loss is 0.67 dB. The 3-dB notch band is from 4.78 to 5.6 GHz centered at 5.19 GHz with an improved notch depth of 32.3 dB. The measured notch frequency shifted slightly downward due to fabrication tolerances. The return loss is better than 13 dB in the passband except in the notch region. A maximum group delay variation of 0.42 ns is seen in the passband before the notch and 0.49 ns after the notch.

In filter "B," the notch depth is increased in comparison to filter "A" at the cost of the filter length. To have an optimum solution for the filter length and notch depth, the EBG cells are also coupled to the low impedance line of the MMR filter as shown in Figure 3.7a (Filter "C"). Here, three EBG unit cells are placed on either side of the line at three different locations. The length of the filter is 53.31 mm ($1.473\lambda_g$, λ_g at 5.19 GHz). The measured and simulated results are shown in Figure 3.7b. The measured 3-dB passband is from 3.3 to 9.99 GHz with a fractional bandwidth of 100%. The measured and simulated insertion losses are 1.26 dB and 0.41 dB, respectively, at the center frequency (6.63 GHz). The measured 3-dB notch is from 4.69 to 5.55 GHz centered at 5.16 GHz with a depth of 18.8 dB.

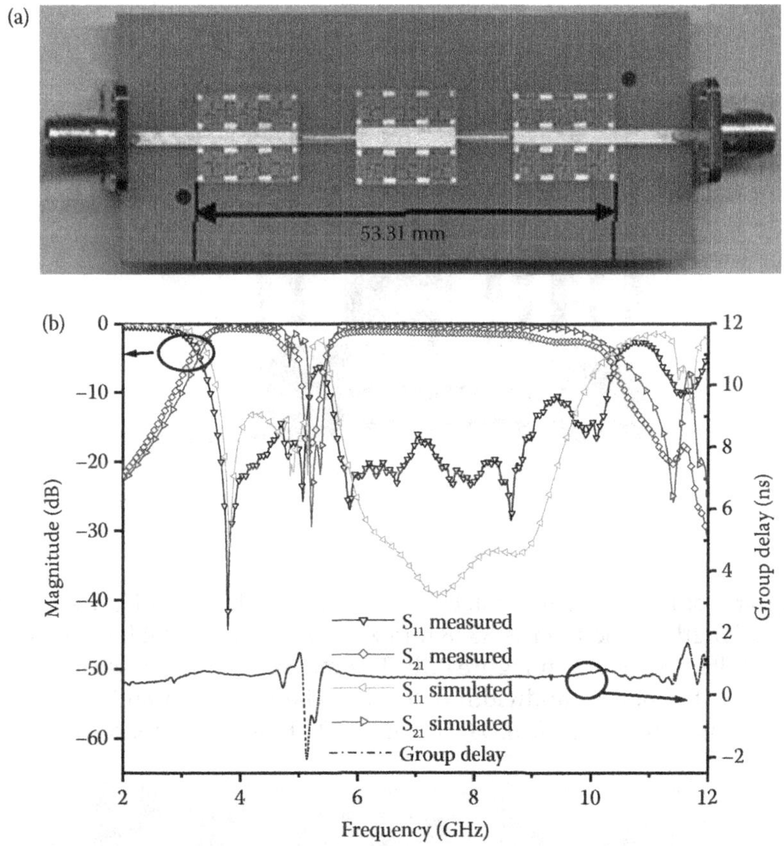

FIGURE 3.7
Band-notched UWB filter with EBG structure coupled to input line, output line, and low impedance line of the filter (Filter C): (a) photograph and (b) measured and simulated results.

The return loss hovers around 10 dB throughout the passband except in the notch region. The maximum group delay variation in the passband is 0.34 ns before the notch and 0.38 ns after the notch. Thus filter "C" is the optimal solution, in terms of length and depth of the notch, amongst the three configurations studied.

3.3.2 Dual Band-Notched UWB Filter

The notch frequency depends on the effective inductance and capacitance of the unit cell. By changing the size of the unit cell, the meander line and interdigital capacitor lengths are varied resulting in an overall change in the inductance and capacitance. Thus, the resonant frequency of the EBG cell can be varied. To have dual notch bands, two differently sized EBG cells are used. An EBG with a unit cell size of 4.32 mm × 4.32 mm is used in all the previous filter circuits. In this dual band-notched UWB filter, seven unit cells sized 4.32 mm × 4.32 mm (Figure 2.1) placed on either side of the output line are responsible for the first notch, and seven EBG cells sized 3.65 mm × 3.65 mm placed on either side of the input line are responsible for the second notch. An EBG cell, which is 3.65 mm × 3.65 mm, is shown in Figure 3.8; it is arrived at by decreasing the size of the unit cell in Figure 2.1, which results in a smaller meander line and smaller interdigital capacitor, making the notch frequency shift upward.

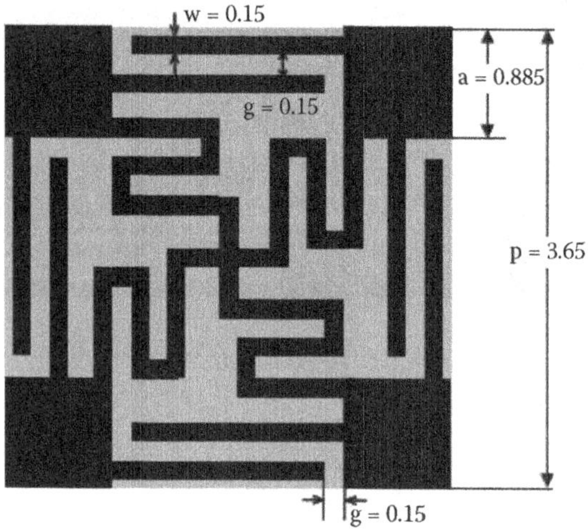

FIGURE 3.8
Unit cell sized to 3.65 mm × 3.65 mm.

A photograph of the dual band-notched UWB filter is shown in Figure 3.9a (Filter "D"). The effective length of the filter is 83.18 mm ($2.298\lambda_g$, λ_g at 5.19 GHz). The measured and simulated results are shown in Figure 3.9b. The measured 3-dB passband is from 3.43 to 9.84 GHz with a fractional bandwidth of 96.61%. The measured and simulated insertion losses at the center frequency (6.635 GHz) are 1.59 dB and 0.65 dB, respectively. The first

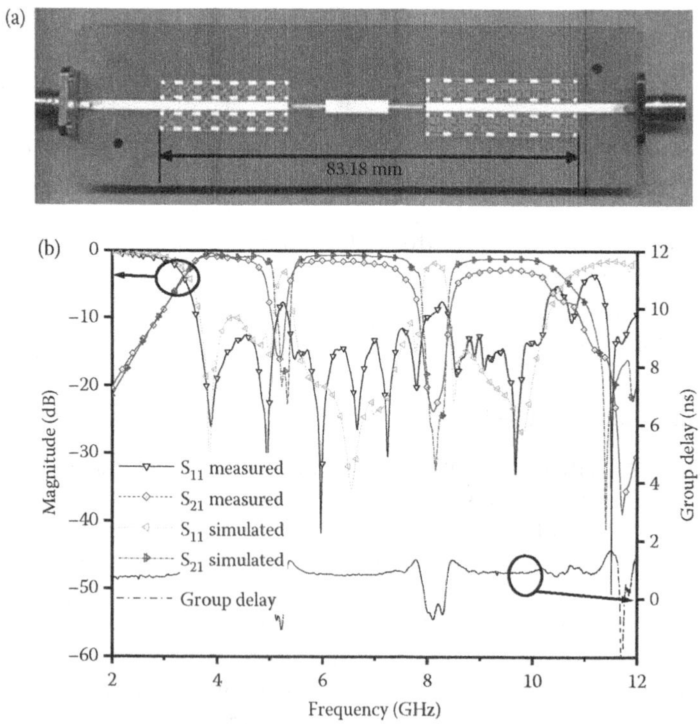

FIGURE 3.9
Dual band-notched UWB filter (Filter D): (a) photograph and (b) measured results.

TABLE 3.1

Summary of Characteristics of Band-Notched UWB Filters

Structure	3-dB Notch Band (GHz)	Notch Depth (dB)	Total Filter Length (λ_g at 5.19 GHz)	Maximum Group Delay in Passband
Filter "A" (Figure 3.4a)	4.88–5.48	16.8	$1.592\lambda_g$	0.54 ns
Filter "B" (Figure 3.6a)	4.78–5.6	32.3	$2.427\lambda_g$	0.49 ns
Filter "C" (Figure 3.7a)	4.69–5.55	18.8	$1.473\lambda_g$	0.38 ns
Dual Band-Notched UWB Filter. Filter "D" (Figure 3.9a)	4.85–5.46 7.45–9.03	16 23	$2.298\lambda_g$	0.4 ns

3-dB notch band is from 4.85 to 5.46 GHz centered at 5.16 GHz with a notch depth of 16 dB. The second 3-dB notch band is from 7.45 to 9.03 GHz centered at 8.24 GHz with a notch depth of 23 dB. The return loss is better than 11.5 dB throughout the passband except in the notch regions. The maximum group delay variation in the passband is 0.23 ns before the first notch, 0.4 ns in between the notches, and 0.15 ns after the second notch.

The important characteristics of various band-notched UWB filters studied are summarized in Table 3.1.

3.4 Band-Notched UWB Antenna

A circular monopole antenna [112] is used as a basic UWB antenna in this section. The UWB antenna is designed on a substrate with a relative permittivity of 3.2 and thickness of 0.762 mm. A schematic diagram and photograph of the band-notched UWB antenna are shown in Figures 3.10 and 3.11, respectively. It consists of circular disc with a radius of 7.86 mm on the front side of the substrate, connected to a 50 Ω microstrip transmission line with a width of 1.848 mm. A gap g of 0.06 mm is maintained between the circular disc

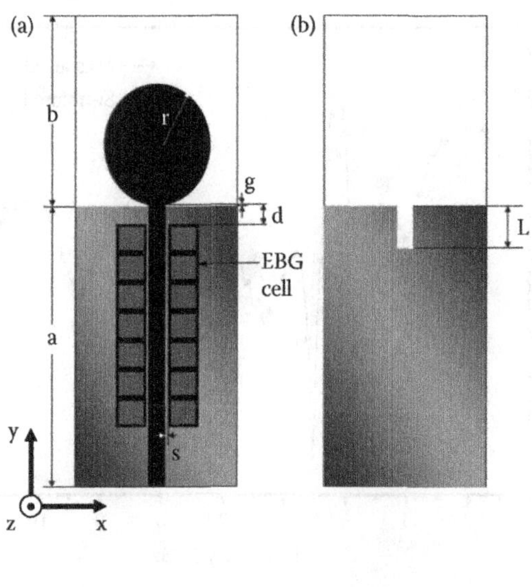

FIGURE 3.10
Schematic diagram of the band-notched UWB antenna: (a) front view and (b) back view. (a = 44.45, b = 20.55, s = 0.1, d = 1, g = 0.06, L = 4.95, r = 7.86, and all dimensions are in mm.)

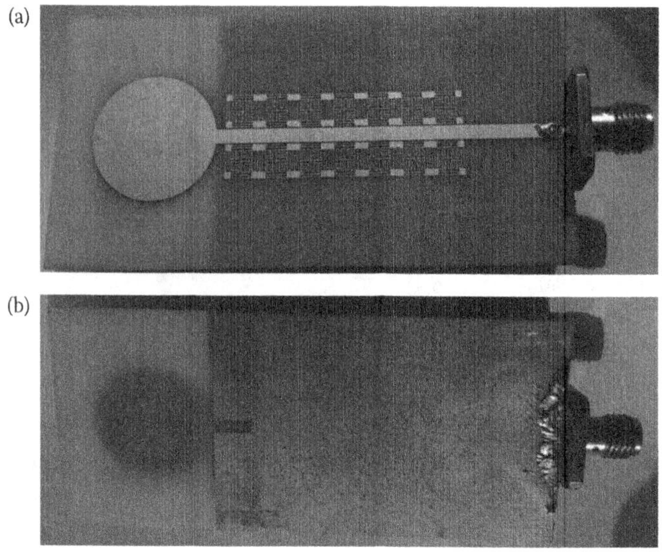

FIGURE 3.11
Photograph of the band-notched UWB antenna: (a) front view and (b) back view.

and the ground plane on the back side of the substrate. A slot of 4.95 mm × 1.848 mm is cut in the ground plane just below the feed line for broadband matching. To achieve a notch in the UWB antenna, the notch circuit in filter "A" is used, where seven unit cells (unit cell shown in Figure 2.1) are placed on either side of the feed line with spacing s of 0.1 mm at a distance d of 1 mm from the lower edge of the circular disc.

$|S_{11}|$ of the antenna is measured using the ZVB20 Rohde & Schwarz vector network analyzer. The measured and simulated $|S_{11}|$ of the band-notched UWB antenna are plotted in Figure 3.12.

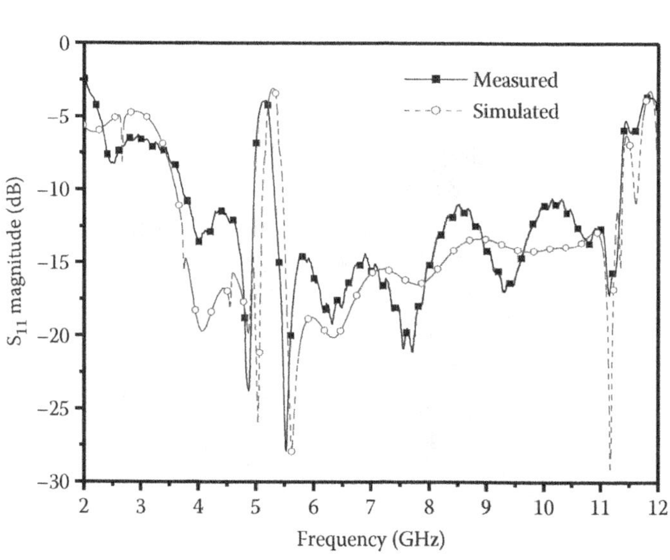

FIGURE 3.12
Measured and simulated $|S_{11}|$ of the band-notched UWB antenna.

The measured 10-dB bandwidth of the band-notched UWB antenna is from 3.73 to 11.35 GHz with a notch band from 4.96 to 5.32 GHz. The measured notch frequency is shifted slightly downward compared to the simulated result due to fabrication errors.

Radiation pattern measurements are done at 5.2 GHz, 6.85 GHz, and 8.5 GHz in the anechoic chamber facility at CARE, IIT Delhi. The measured, as well as simulated, E-plane (y–z plane) patterns are shown in Figure 3.13, and the H-plane (x–z plane) patterns are shown in Figure 3.14. Measured radiation patterns match well with the simulated patterns. The radiation pattern is omnidirectional in the H-plane as expected.

Figure 3.15 shows the gain of the UWB antenna and the band-notched UWB antenna. The gain of the band-notched UWB antenna at notch frequency (5.2 GHz) is decreased by 16 dB when compared to the antenna without a notch as shown in Figure 3.15. Since the return loss at 5.25 GHz is only 5.7 dB, it is not very close to 0 dB, thus the band-notched

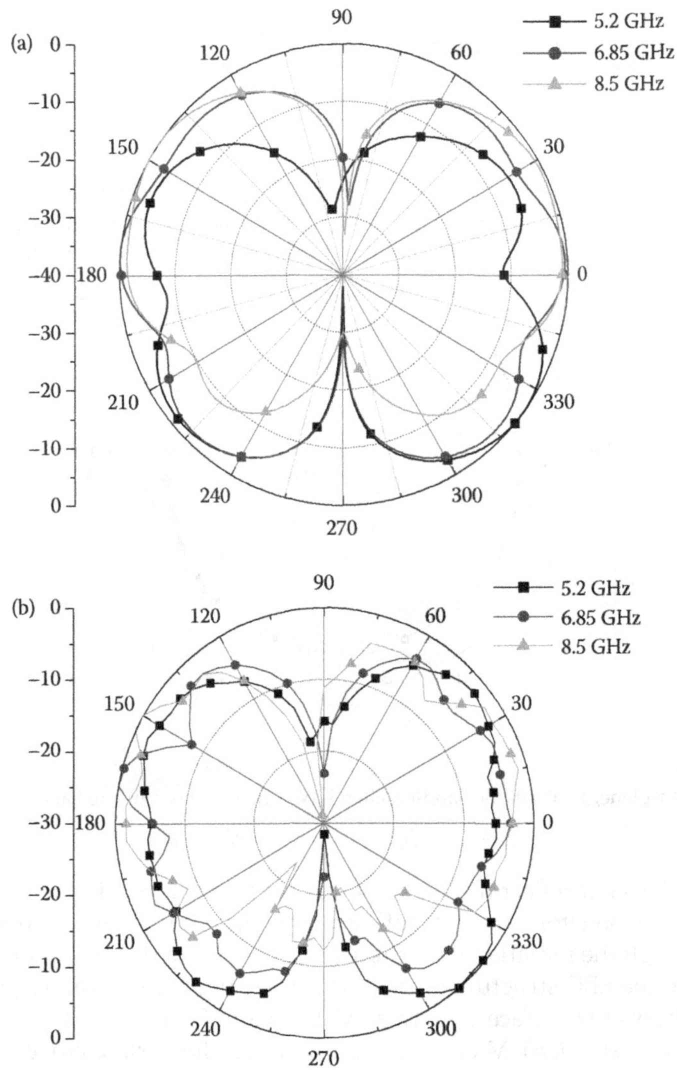

FIGURE 3.13
E-plane pattern (y–z plane, φ = 90°) of the band-notched UWB antenna: (a) simulated and (b) measured.

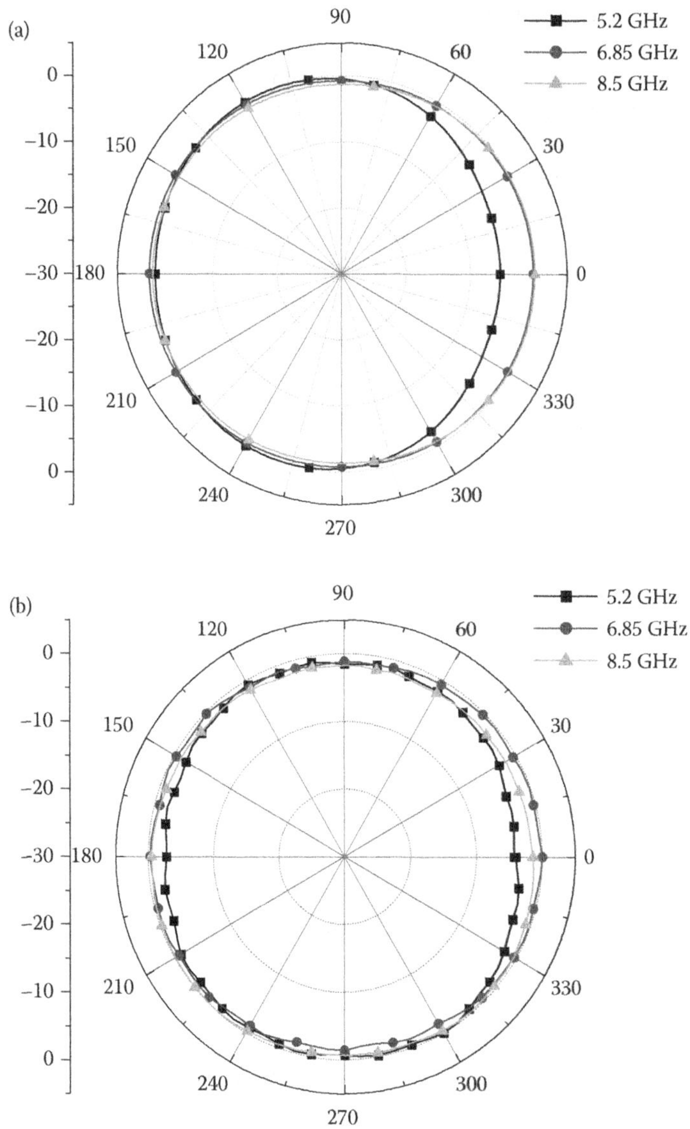

FIGURE 3.14
H-plane pattern (x–z plane, $\varphi = 0°$) of the band-notched UWB antenna: (a) simulated and (b) measured.

UWB antenna does not reflect the entire power of the notch band. The remaining energy is confined in the resonator, thus presenting a large difference in the gain measured in the notch band, though the radiation patterns are smooth. From the $|S_{11}|$ plot and gain plot, it can be seen that the EBG structure is effectively rejecting the WLAN frequency.

Figure 3.16 shows the surface current at 5.2 GHz and 8 GHz in the band-notched UWB antenna obtained using CST Microwave Studio. From the surface current plots, it can be seen that electromagnetic energy is coupled with the EBG cells. At 5.2 GHz, in the notch band where the EBG structure is under resonance, the coupled energy is concentrated in the EBG structure. More concentrated energy is seen in the first EBG cell, and gradually

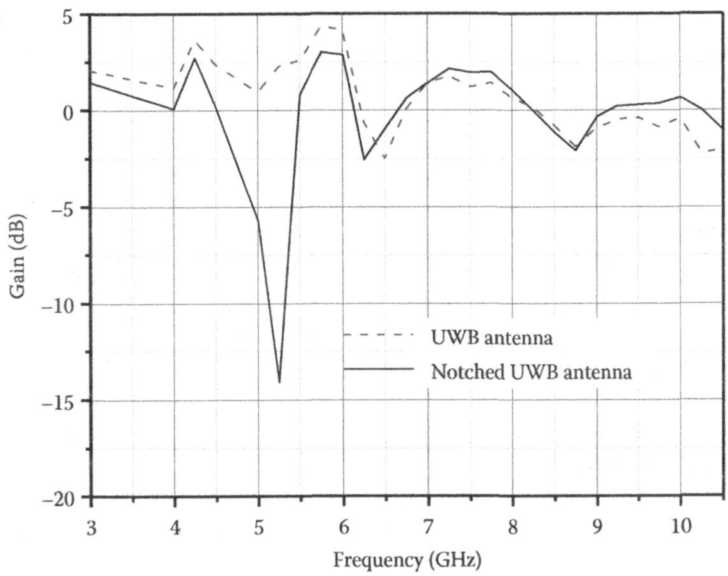

FIGURE 3.15
Gain of the UWB antenna and the band-notched UWB antenna.

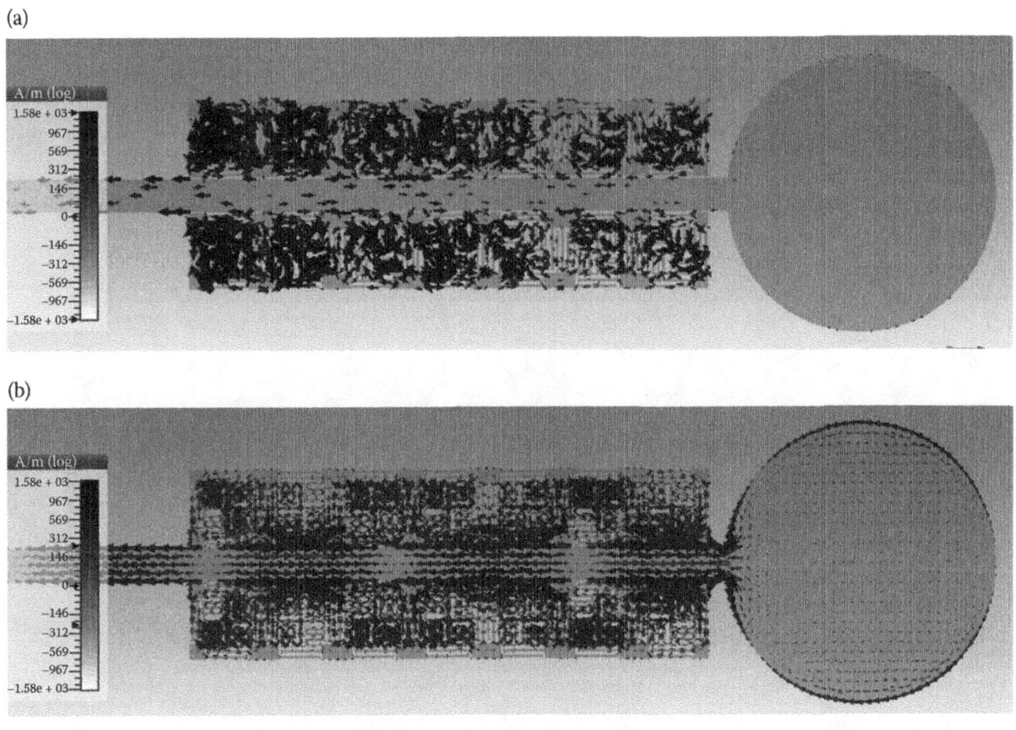

FIGURE 3.16
Surface current in the band-notched UWB antenna at (a) 5.2 GHz and (b) 8 GHz.

decreases in the subsequent cells. At 8 GHz, the EBG is not resonating, the coupled energy is back coupled to the line, and the wave propagates in a forward direction to the antenna.

3.5 Conclusion

Single band-notched UWB filters are developed and tested successfully. All three variants give a notch band around 5.19 GHz. Filter "B" has a good rejection depth of 32.3 dB when compared to Filter "A" (depth of 16.8 dB), but the length of the filter is higher. Filter "C" appears to be the optimal solution, with a smaller overall length and better rejection depth of 18.8 dB. A dual band-notched UWB filter is developed with a notch band around 5.16 and 8.24 GHz. Along with the filter application, in a similar manner, the EBG notch filter is also applied to a UWB antenna.

In Chapter 4, the notch band in the UWB filter is reconfigured by modifying the length of each unit cell by connecting an additional structure using PIN diode switches.

4

Reconfigurable Periodic Structures

4.1 Introduction

Reconfigurable EBG structures are very useful as their characteristics can be changed electronically. Various reconfigurable EBG structures and their applications are reported in References 27 and 78–89. In this chapter, a reconfigurable planar EBG structure is developed by changing its effective capacitance and is further applied to create a reconfigurable notch in a UWB filter.

Several reconfigurable band-notched UWB filters are reported in the literature. These include the reconfigurable notch (ON/OFF) in UWB filter [113–116] and the stepped impedance resonator (SIR) used for reconfiguration of a notch from one position to another in Reference 75, whereas having a via in the structure makes fabrication difficult. Although in Reference 116 a planar structure is used to obtain a reconfigurable notch, the passband insertion loss is poor. EBG structures are also used for reconfigurable bandstop filters in microstrip technology [27,89], but the structure in Reference 27 does not result in a smooth passband. The structure reported in Reference 89 gives a smooth passband, but it also requires conducting vias.

In this chapter, the fixed band-notched UWB filter presented in the previous chapter is made reconfigurable. The EBG structure used here does not require vias and has a smooth passband. The equivalent circuit of the reconfigurable unit cell in the ON and OFF states coupled to a transmission line is developed. A bandwidth reconfigurable notch filter is also developed utilizing another configuration of the unit cell.

4.2 Switchable Band-Notched UWB Filter

As seen in the previous chapter, the notch frequency depends on the effective inductance and capacitance of the unit cell. In order to change the notch frequency, an additional capacitance is added to the unit cell by connecting an additional structure using PIN diode switches. Figure 4.1 shows the metal pattern of the unit cell and the additional section. The size of the unit cell is 4.12 mm × 4.12 mm. The additional structure is predominantly capacitive and sized 1.5 mm × 4.12 mm. The unit cell and additional section are separated by a 0.3 mm gap for placing a PIN diode. The metal pattern shown in Figure 4.1 is on top of the conductor-backed substrate with $\varepsilon_r = 3.2$ and a thickness 0.762 mm.

By placing a switch (a PIN diode or a varactor) between the unit cell and the additional section, the overall size of the unit cell is increased with a switch in the ON state, thereby

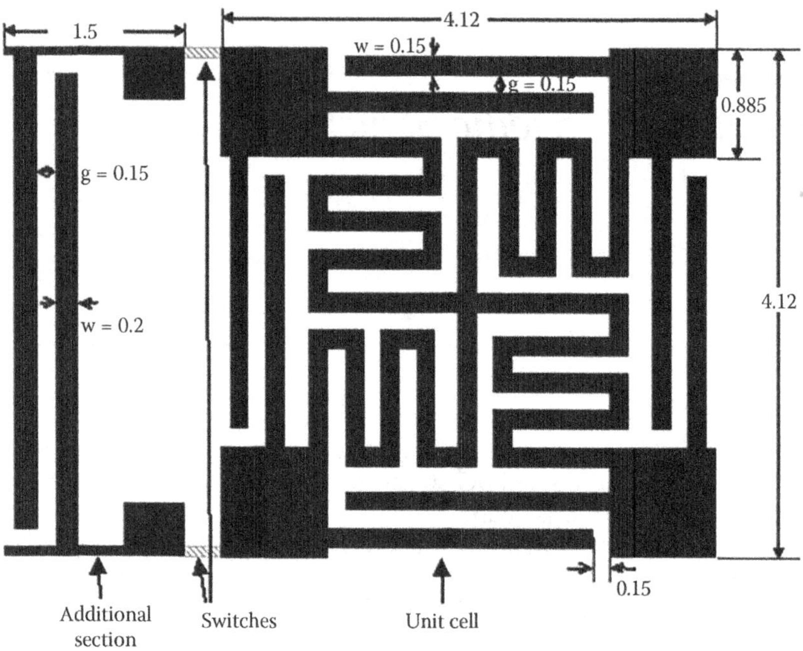

FIGURE 4.1
Schematic diagram of metal pattern of the reconfigurable EBG. (All dimensions in mm.)

the effective capacitance is increased and the notch frequency shifts down. The one-dimensional array of a modified structure is capacitively coupled to the input/output lines of MMR UWB filter on either side as shown in Figure 4.2a. The zoomed view of the EBG cells with switches and biasing circuitry is shown in Figure 4.2b. Fourteen unit cells of EBG (7 on either side of the input/output lines) are used. Sixteen PIN diodes (MA4SPS402) are used to modify the 14 EBG cells. Since EBG structures are periodic structures, a single diode cannot change the geometry of all the unit cells; therefore, more diodes are required for reconfigurability of EBG structures.

The concept of switching is first verified using ideal ON/OFF conditions of the diode. Filter prototypes in the ideal ON state and ideal OFF states are fabricated, and their photographs are shown in Figure 4.3a and b, respectively.

The inset in Figure 4.3a shows a zoomed view of the unit cell connected to an additional section by a 0.15 mm wide metal strip used for realizing the ON state. The width of this metal does not show any significant effect. The only 10 MHz difference is observed when simulations are performed with metal strips with widths of 0.432 and 0.15 mm. The OFF state is realized by removing these strips which can be seen in the zoomed view in the inset of Figure 4.3b. Figure 4.4a and b shows the simulated and measured results of the filter in ideal ON and OFF states, respectively. The biasing circuit that would be required for use with an actual diode is also fabricated. However, no components are used while studying ideal cases. As observed from the results, the measured notch frequency is slightly lower than the simulated values because of the fabrication errors. Slight over-etching makes the lines in the unit cells narrower which increases the effective inductance. This results in slightly lower notch frequency than expected.

An inspection of the fabricated structure under a microscope reveals that, on average, over-etching of about 20 μm has taken place. To confirm the effect of over-etching,

(a)

(b)

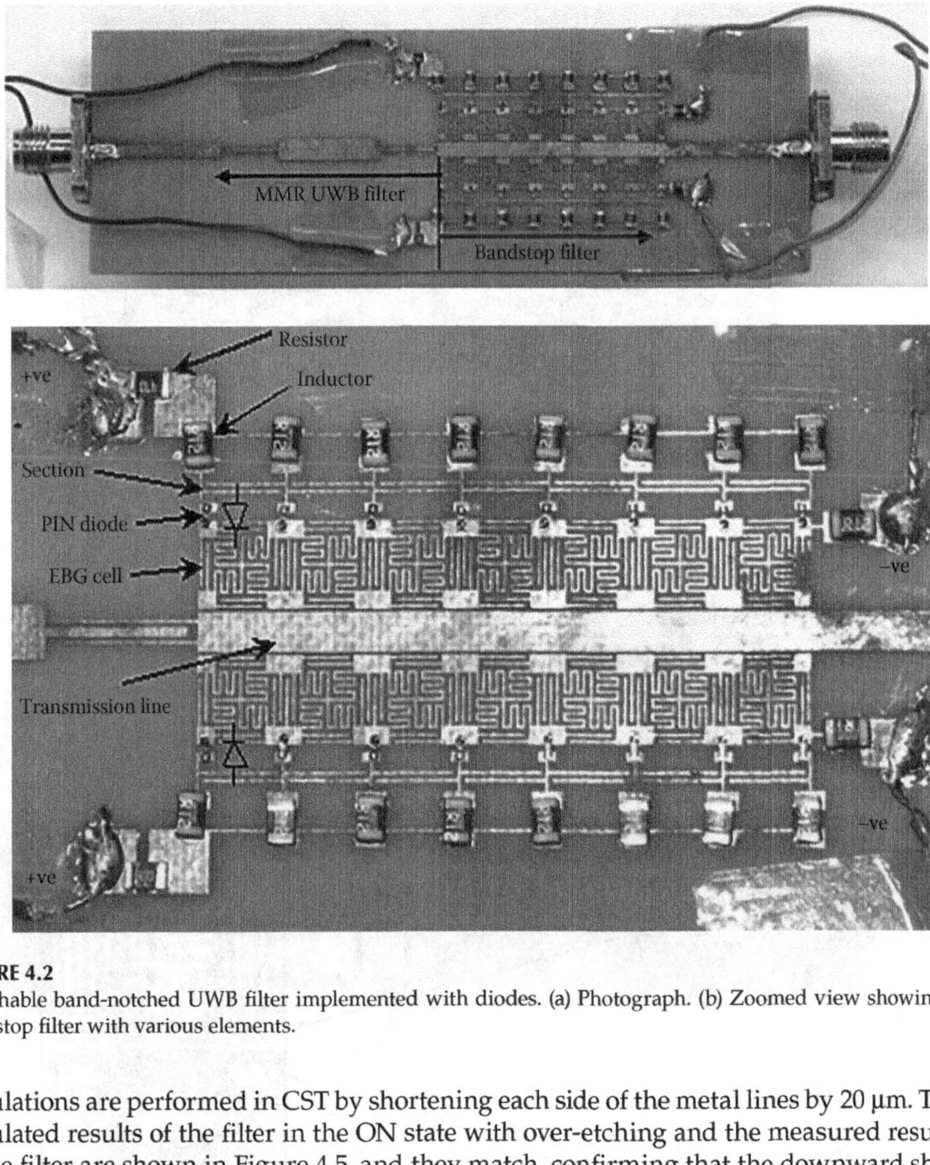

FIGURE 4.2
Switchable band-notched UWB filter implemented with diodes. (a) Photograph. (b) Zoomed view showing a bandstop filter with various elements.

simulations are performed in CST by shortening each side of the metal lines by 20 μm. The simulated results of the filter in the ON state with over-etching and the measured results of the filter are shown in Figure 4.5, and they match, confirming that the downward shift in the measured results is due to over-etching during the fabrication process.

A substrate parameter deviation studied is performed; it is observed that the notch band is shifting down with an increase in the dielectric constant.

MA-COM PIN diodes (MA4SPS402) are used as switches to connect the additional sections with the unit cells. In order to bias the diodes, 120 nH chip inductors are used as RF chokes (Figure 4.2b). To simulate the structure with diodes and inductors, initially full wave simulation of the filter is done with one discrete port in place of each diode and each inductor along with input and output ports in CST microwave studio. The simulated S-parameter file with 34 ports (.s34p) is then imported into ADS software. The diode model and inductor's S-parameter file are used to complete the simulation setup. The CST-simulated results with ideal conditions and ADS-simulated results with diodes in place are compared in Figure 4.6a and b for the ON and OFF states, respectively. Because of

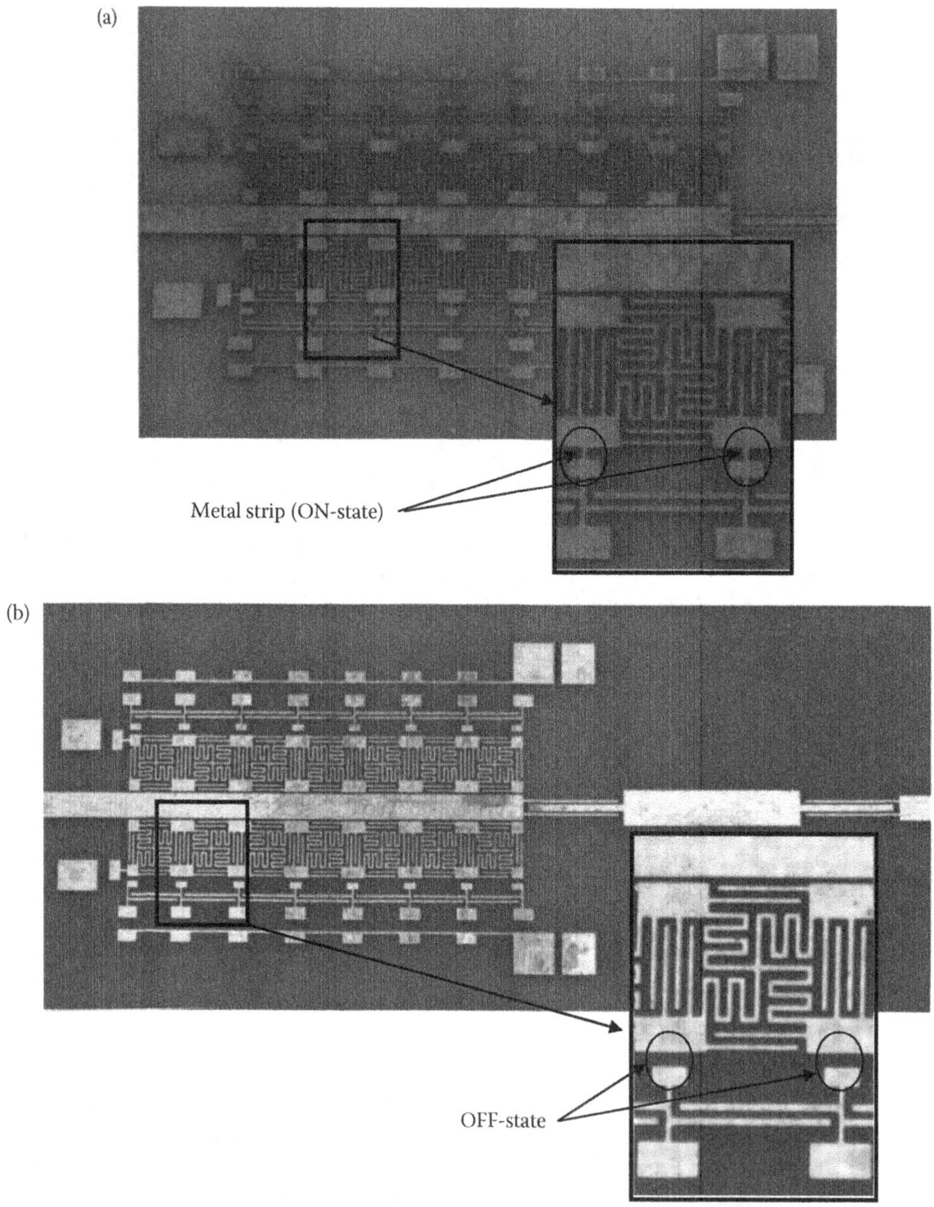

FIGURE 4.3
Photograph of the switchable band-notched UWB filter. (a) Ideal ON state. (b) Ideal OFF state.

the diode capacitance, the ADS results are slightly shifted downward compared to CST-simulated results under ideal conditions.

All 16 PIN diodes in the switchable band-notched UWB filter (Figure 4.2) are connected in a parallel and biased with a total source current of 160 mA in the ON state. The measured results of the switchable band-notched UWB filter with diodes in ON and OFF states are shown in Figure 4.7. In the diode OFF state, the 10-dB notch bandwidth is 0.316 GHz centered around 5.545 GHz, and in the ON state, it is 0.344 GHz centered around 5.095 GHz.

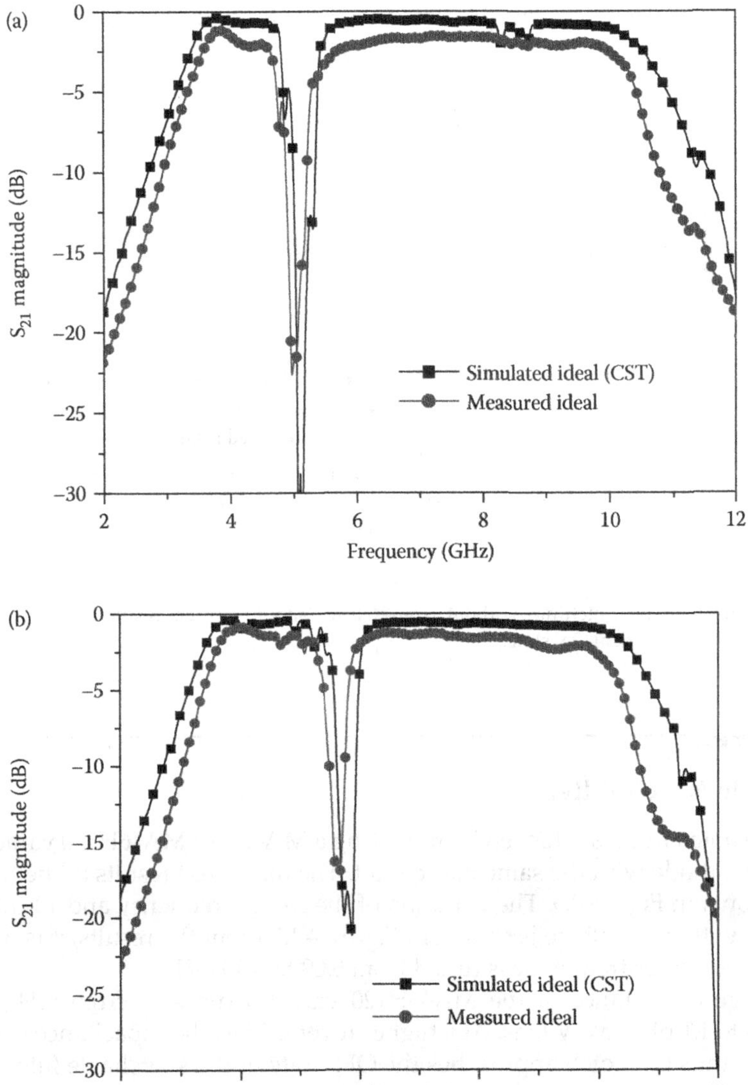

FIGURE 4.4
Simulated and measured results of the switchable band-notched UWB filter. (a) Ideal ON state. (b) Ideal OFF state.

A 450 MHz shift in the frequency is observed by switching the diodes. The insertion loss at 6.825 GHz in both states is 1.52 dB. The notch depth is better than 18 dB in both states. As the EBG cells are made up of narrow lines, they tend to radiate, thus the reflection coefficient ($|S_{11}|$) in the notch band is not close to 0 dB. However, the transmission coefficient in the notch band is low, which is desired in both the states. Isolation of 16 dB is achieved between the ON to OFF states. The measured results with diodes show a downward shift in frequency compared to ADS-simulated results, as shown in Figure 4.8, due to fabrication imperfections.

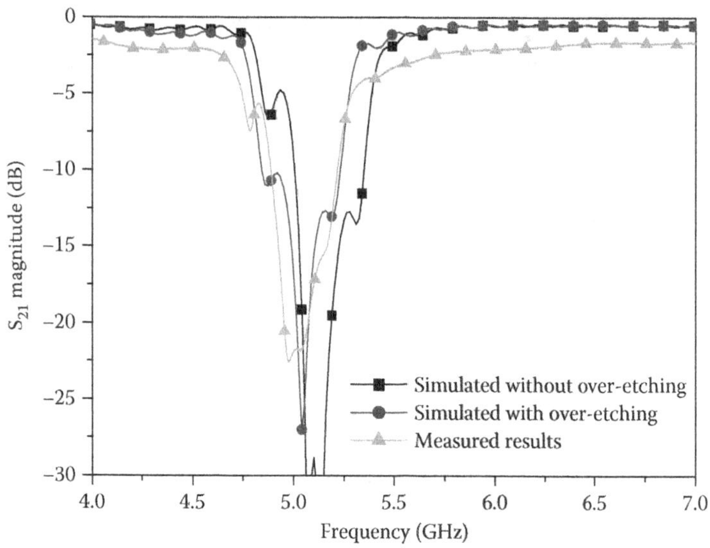

FIGURE 4.5
Simulated results with over-etching, without over-etching and measured results of the switchable band-notched UWB filter in the ideal ON state.

4.3 Tunable Notch Filter

The tunable notch filter is achieved by mounting a MA-COM MA46H120 varactor diode in place of a PIN diode with the same bias circuit. The measured results of the tunable notch filter are shown in Figure 4.9. The variation of the center frequency and 10-dB bandwidth of the notch with bias voltage is shown in Figure 4.10. From the results, it is observed that the notch band center frequency is tuned from 5.09 to 5.43 GHz.

The average capacitance of the MA46H120 varactor diode is from 0.14 pF at a 15 V reverse bias to 1.1 pF at a 0 V bias. At a higher reverse bias, the capacitance of the varactor diode is low, thus the notch approaches the OFF state of the switchable filter using a PIN diode. At zero bias, the capacitance is large, therefore the notch shifts toward the ON state of the switchable filter. A higher ON-to-OFF state capacitance varactor than in MA46H120 can give tuning ranges close to a switchable case.

A 450 MHz shift is achieved in a switchable case and a 340 MHz tuning is achieved in a tunable filter case. In order to change this switchable/tunable frequency range, the additional section connected to the unit cell needs to be modified.

4.4 Equivalent Circuit of Frequency Reconfigurable Unit Cell

A single reconfigurable unit cell coupled to the microstrip transmission line is shown in Figure 4.11. It consists of a microstrip line coupled with the unit cell sized 4.12 mm × 4.12 mm.

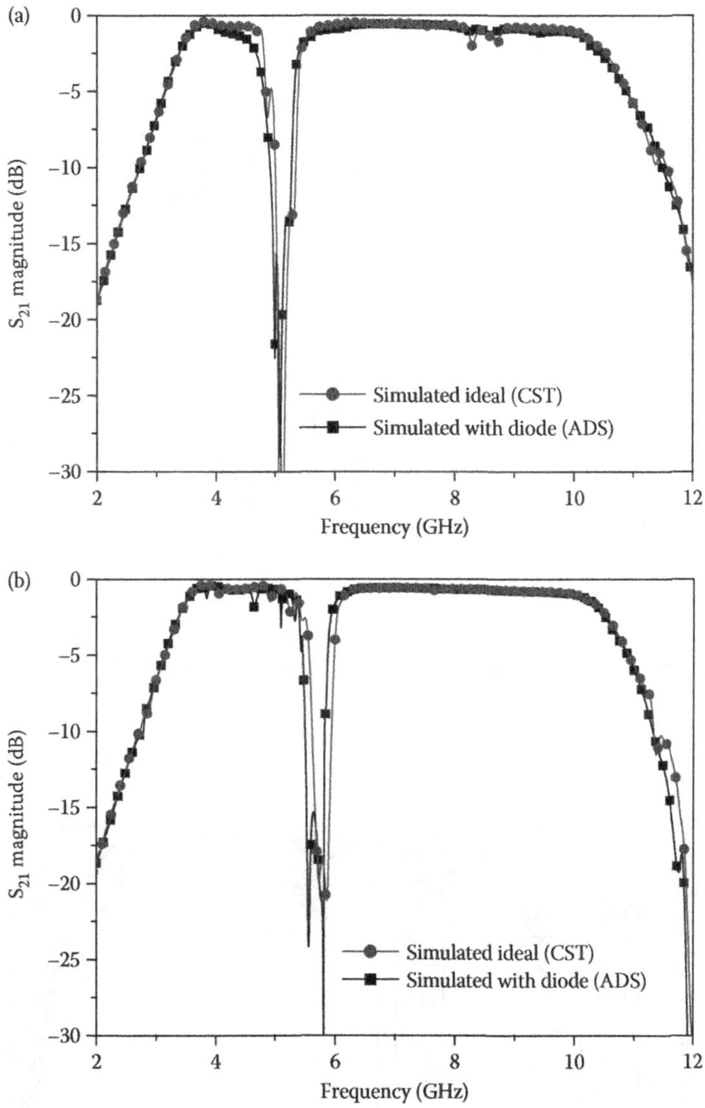

FIGURE 4.6
Simulated results of switchable band-notched UWB filter in CST and ADS. (a) ON state. (b) OFF state.

This unit cell is connected to an additional section by means of a small metal strip with a width of 0.1 mm.

Following the procedure presented in Section 2.6, the equivalent circuit model of the unit cell and the coupling (between the line and the unit cell) are developed. The additional section consists of a two-finger-interdigital capacitor connected to the small pads by a narrow line. The narrow line is represented by an inductor and the small pad is represented by a shunt capacitor arrived at by a parallel plate capacitance equation. The equivalent circuit model of the circuit shown in Figure 4.11 is derived and the same is shown in Figure 4.12 with parameters: $L = 2.2034$ nH, $C_L = 4.58$ fF, $C = 41.1$ fF, $L_C = 1.073$ nH, $C_S = 0.216$ pF,

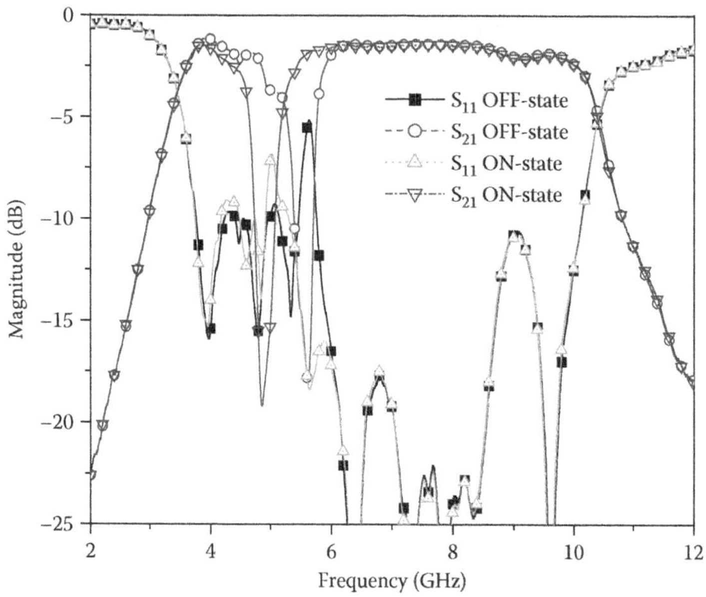

FIGURE 4.7
Measured results of the switchable band-notched UWB filter with diodes in the ON state and OFF states.

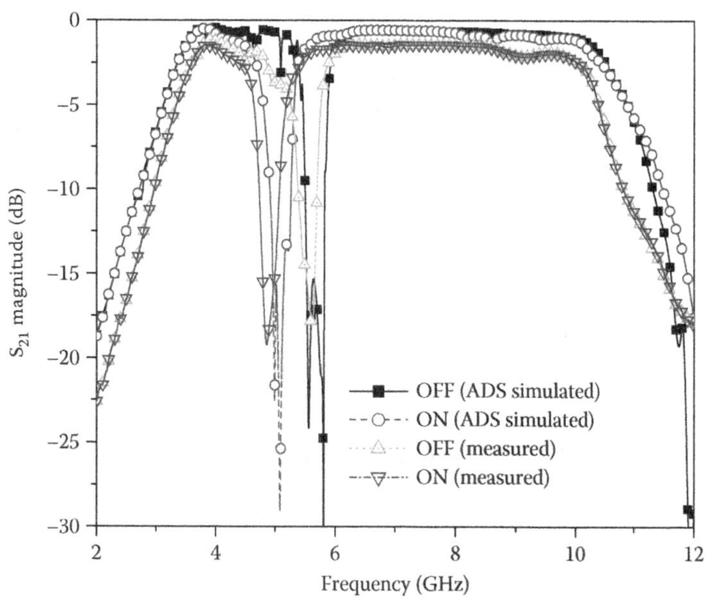

FIGURE 4.8
Comparison of measured and ADS-simulated results of the switchable band-notched UWB filter implemented with diodes.

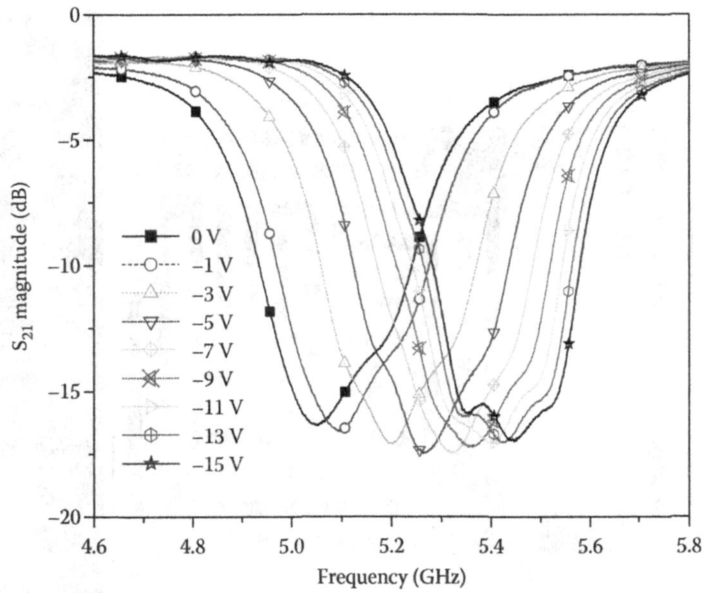

FIGURE 4.9
Measured results of the tunable notch filter with a varactor diode in reverse bias.

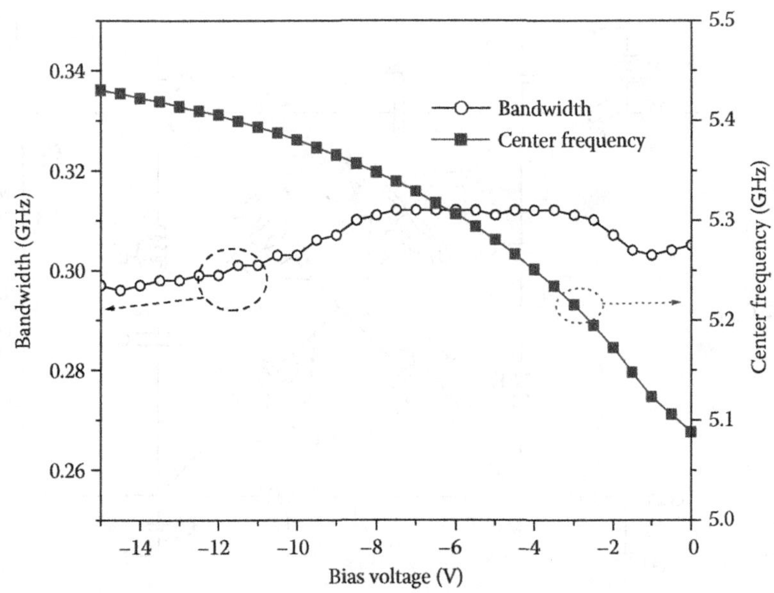

FIGURE 4.10
Variation of frequency and bandwidth of the tunable filter with bias voltage of varactor diode.

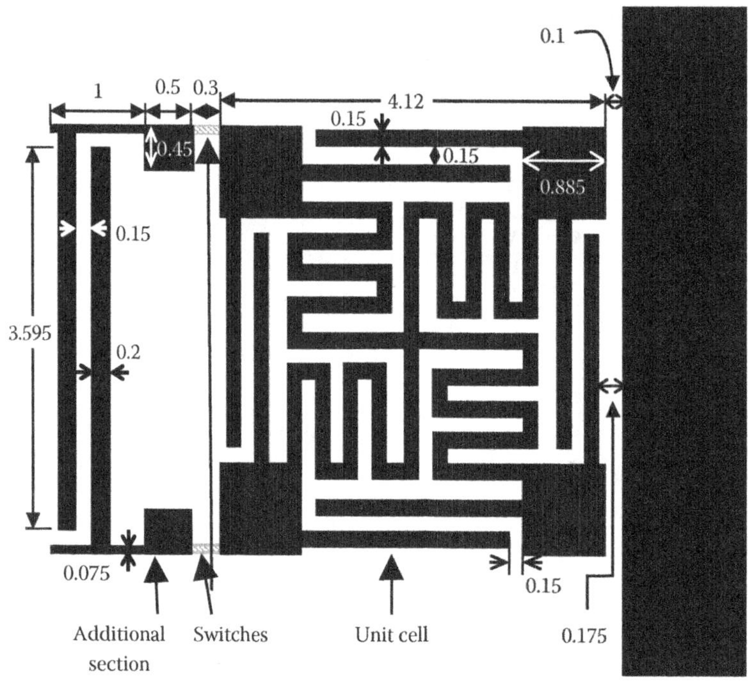

FIGURE 4.11
Reconfigurable EBG unit cell coupled to a transmission line.

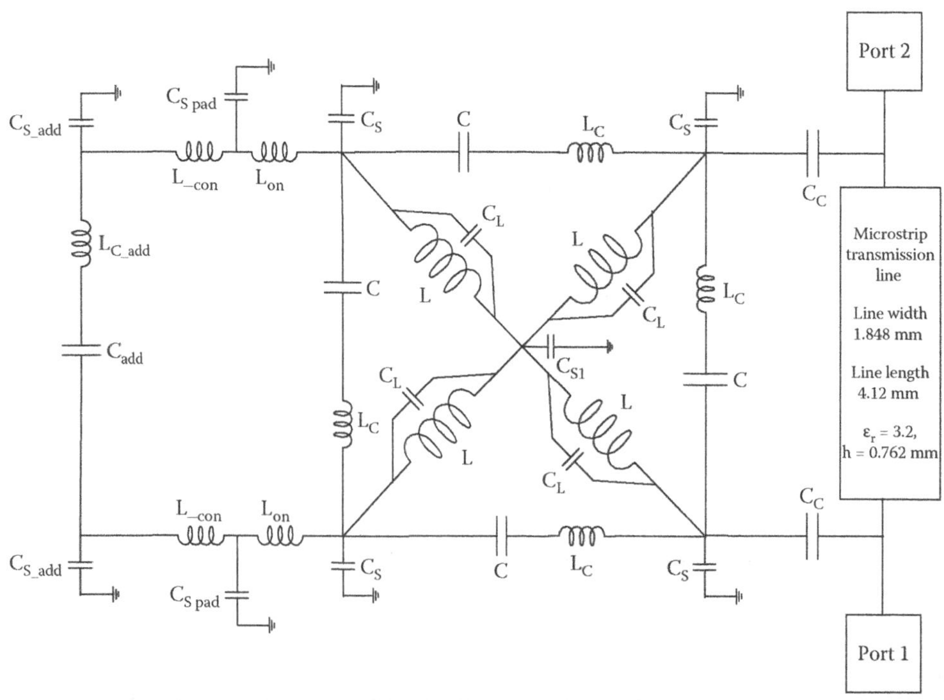

FIGURE 4.12
Equivalent circuit of the reconfigurable EBG unit cell coupled to a transmission line.

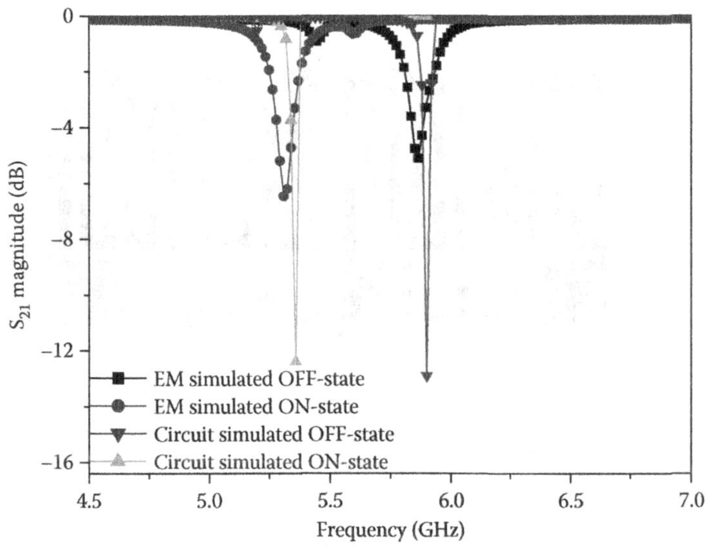

FIGURE 4.13
Comparison of circuit-simulated and EM-simulated results of the reconfigurable unit cell coupled to a transmission line.

$C_{S1} = 0.212$ pF, $C_{add} = 80.5$ fF, $L_{C\_add} = 1.741$ nH, $C_{S\_add} = 111.5$ fF, $L_{\_con} = 0.76$ nH, $C_{Spad} = 26.4$ fF, $L_{on} = 0.16$ nH, and $C_C = 0.051$ pF.

Figure 4.13 shows the comparison of circuit-simulated results and EM-simulated results. In the OFF state, there is no connection between the unit cell and the additional section. In the ON state, a metal strip is used as connector between the unit cell and the additional section, the metal strip is represented by an inductor L_{on} in the equivalent circuit. Circuit-simulated results match well with EM-simulated results, and a difference of only 50 MHz is observed in the ON state and 30 MHz in the OFF state.

4.5 Bandwidth Reconfigurable Bandstop Filter

The bandwidth reconfigurable bandstop filter is one in which the bandwidth of the bandgap is changed electronically. It is seen in Figure 4.10 that bandwidth does not change much in the previous structure; hence a new structure is proposed for bandwidth reconfiguration.

The metal pattern shown in Figure 4.14 is printed on a conductor-backed substrate with $\varepsilon_r = 3.2$ and thickness $h = 0.762$. The unit cell is shown in the dotted line. The meander line inductor and interdigital capacitor are connected between two small pads. In the inductor and capacitor, the width of each line is 0.15 mm and the gap between lines is 0.15 mm. As the inductor and capacitor are connected as a parallel, thus the equivalent circuit of the unit cell is a parallel LC circuit. The resonance frequency and bandwidth are proportional to the effective inductance and capacitance given by Equations 1.1 and 1.2.

Figure 4.14 shows two unit cells connected by a switch. The current through the switch provides the inductance between the cells. From Equations 1.1 and 1.2, it can be seen that as the inductance increases bandwidth of the stopband increases and the center frequency

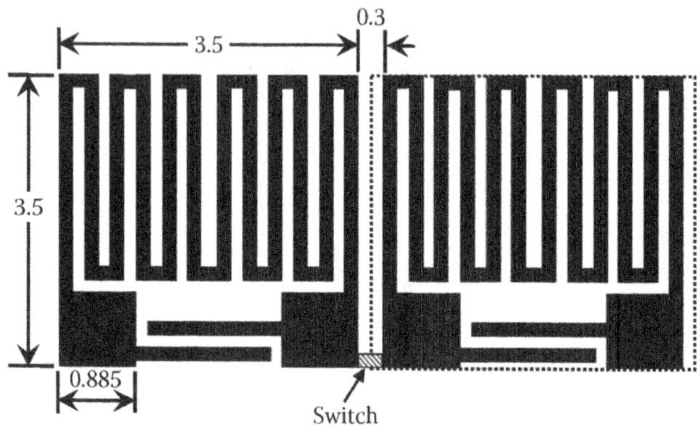

FIGURE 4.14
Two unit cells connected by a switch.

decreases. Thus, connecting unit cells to each other or disconnecting them can increase or decrease the bandwidth, respectively.

As seen in the previous chapters, when the EBG cells are coupled to a microstrip transmission line, it results in a bandstop filter utilizing the bandgap property. Here, seven unit cells shown in Figure 4.14 are coupled to a microstrip transmission line with a gap of 0.1 mm. Figure 4.15 shows the photograph of the fabricated filter and its zoomed view show two EBG cells with diode and bias circuitries.

MA-COM PIN diodes (MA4SPS402) are used to connect EBG cells. Seven EBG cells on each side of the transmission line are connected by 6 PIN diodes placed back to back. 5.6 nH chip inductors from Johanson Technology are used as RF chokes. 1.8 pF capacitors

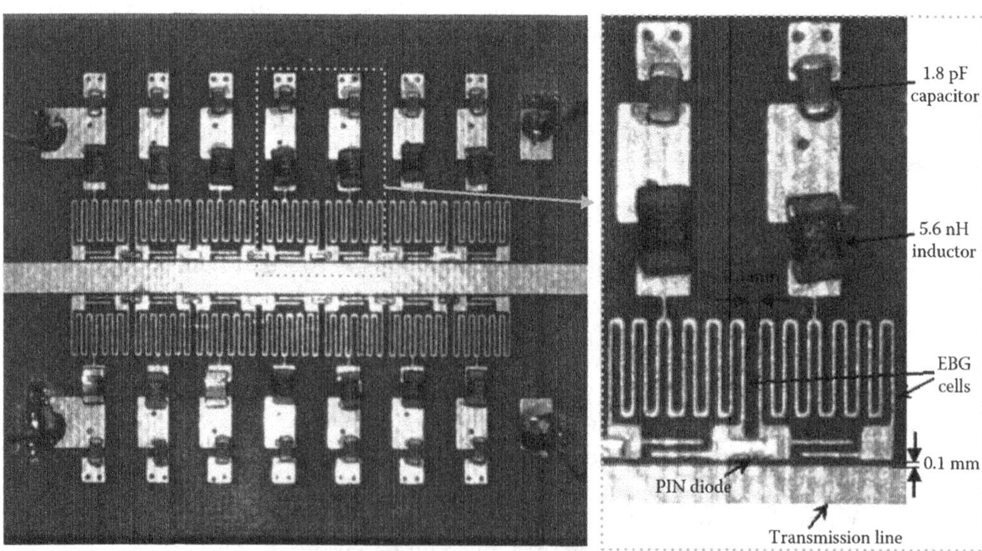

FIGURE 4.15
Photograph of the fabricated bandwidth reconfigurable bandstop filter and its zoomed view showing various elements.

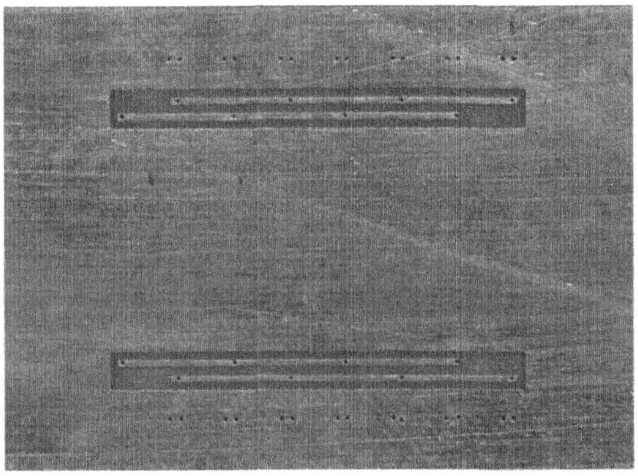

FIGURE 4.16
Photograph of the ground plane of the bandwidth reconfigurable bandstop filter (Figure 4.15).

from the same company are used as bypass capacitors. Since diodes are connected back to back, alternate pads on top of the substrate shown in Figure 4.15 are connected to each other separately on the back side of the substrate through vias as shown in Figure 4.16. A +ve and −ve DC bias voltage is given on alternate pads. All 12 diodes in the circuit are in parallel and biased with a single DC power supply.

Full wave simulations (EM) of the filter are done in a CST Microwave Studio for ideal diode conditions. For the ON state, a metal strip with a width of 0.15 mm is used in between two unit cells for connection, and for the OFF state, the connection between two unit cells is left open. In order to simulate the filter with the diode model, the first full wave simulation is performed in a CST Microwave Studio with one discrete port for each diode along with input and output ports. The next simulations are carried out in ADS software with the S-parameter file from CST (.S_{14p}), the diode model, and S-parameter files of Johanson Technology inductors. The ideal condition EM-simulated results from CST and circuit-simulated results from ADS are shown in Figure 4.17. Because of the diode capacitance in the OFF state, bandwidth of the stopband in ADS simulations is wider than the ideal OFF-state EM simulated results. In order to maintain periodicity, either all the cells have to be connected to each other with diodes in the ON state or all have to be disconnected to each other with diodes in the OFF state.

The measurement of the fabricated structure with diodes is done on an Agilent PNA E8364C. All 12 diodes connected in parallel are biased with a forward current of 180 mA in the ON state.

From the measured results shown in Figure 4.18, the 10-dB bandwidth of the stopband is from 6.84 GHz to 7.47 GHz in the OFF state, and in the ON state, it is from 6.41 GHz to 7.33 GHz. A 290 MHz increase in bandwidth is observed in the ON state as compared to the bandwidth in the OFF state. It is also observed that the center frequency of the stopband is decreased in the ON-state. The reflection coefficient in the stopband is not close to 0 dB because of radiation from EBG cells.

When comparing ADS-simulated results and the measured results from Figures 4.17 and 4.18, it is seen that the measured results have shifted to the lower side because of the fabrication imperfections (over-etching results in thinner lines which increase the inductance, thereby shifting the frequency downward). The measured insertion loss in the passband is 1.93 dB at 8.5 GHz, whereas the simulated insertion loss is 0.5 dB. The insertion

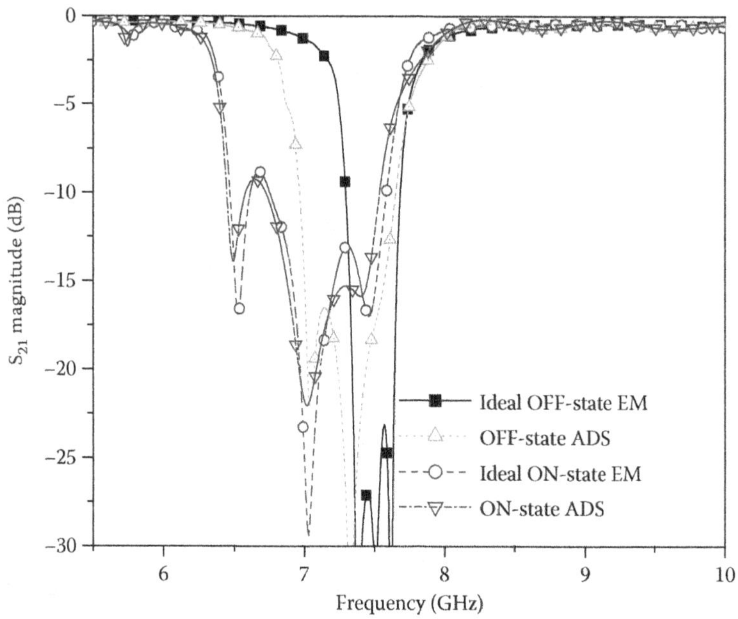

FIGURE 4.17
ADS-simulated results of the bandwidth reconfigurable filter with diode in the ON and OFF states.

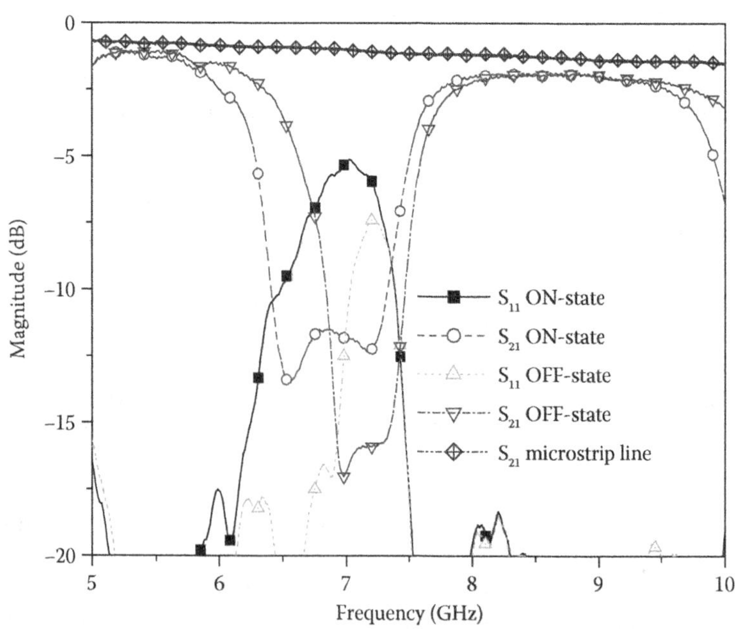

FIGURE 4.18
Measured results of the bandwidth reconfigurable filter with a diode in the ON state, diode OFF-states, and a simple microstrip line.

loss of a simple microstrip line on the same substrate with the same set of connectors is 1.23 dB (Figure 4.18). Hence, the measured loss of 1.93 dB can be caused by the connector losses, substrate losses, and losses due to diodes.

4.6 Conclusion

In this chapter, a reconfigurable unit cell is developed and used for reconfigurable band-notched UWB filter development. Frequency switchable and tunable band-notched UWB filters are implemented using PIN and varactor diodes, respectively. The equivalent circuit of a reconfigurable unit cell coupled to a transmission line is developed in the ON and OFF-states. A bandwidth reconfigurable bandstop filter is also achieved using another configuration of the unit cell.

Up to now, an EBG cell, that is, an edge, coupled to a transmission line is discussed. In the next chapter, the EBG cell is broadside coupled to the transmission line. By using the broadside coupled structure, a harmonic suppressed bandpass filter and diplexer are developed in Chapter 5.

5

Application of Compact Planar Periodic Structure in Bandpass Filters and Diplexers

5.1 Introduction

In this chapter, a bandstop filter (BSF) is developed by coupling the same unit cell reported in the previous chapter to transmission lines in a broadside arrangement. A bandpass filter (BPF) is then created with suppressed harmonics by broadside coupling these unit cells in between two widely separated open-ended lines. Previously, a BPF with spurious free stopband was achieved using electromagnetic bandgap (EBG) resonators in a coplanar waveguide configuration [117] and stepped-impedance resonators [118]. However, split ring resonators have recently been reported to couple two parallel lines to achieve a BPF [119] without harmonic suppression.

Diplexers can split/mix signals with two different frequencies into/from two different channels. Various diplexers are reported using periodic stubs [120], hybrid resonators [121], and composite right/left-handed transmission lines [122]. In this chapter, a novel diplexer is developed using the proposed BPF.

5.2 BSF with Broadside-Coupled EBG Cells

The BSFs reported in the previous chapters consisted of EBG cells and transmission lines in the same plane; thus, the EBG cell is edge coupled to the transmission line. The depth of the stopband depends on the coupling between the transmission line and EBG cell. In this section, EBG cells are placed broadside to the transmission line to increase coupling.

An electric field diagram of a microstrip transmission line is shown in Figure 5.1; electric field lines lie between the line on the top layer and the ground plane on the bottom layer. Figure 5.2 shows a broadside-coupled EBG structure and an edge-coupled EBG structure. The broadside-coupled EBG cell is located in the middle layer between the top line and the ground, and the edge-coupled EBG cell is placed in the same plane as the line. More electric field lines pass through the broadside-coupled EBG cell than through the edge-coupled EBG cell, which means the broadside arrangement permits more coupling. In broadside coupling, the coupling coefficient depends on the height of the dielectric substrate and the overlap between the top and bottom layers. A thin (0.127 mm) substrate is therefore used to increase coupling.

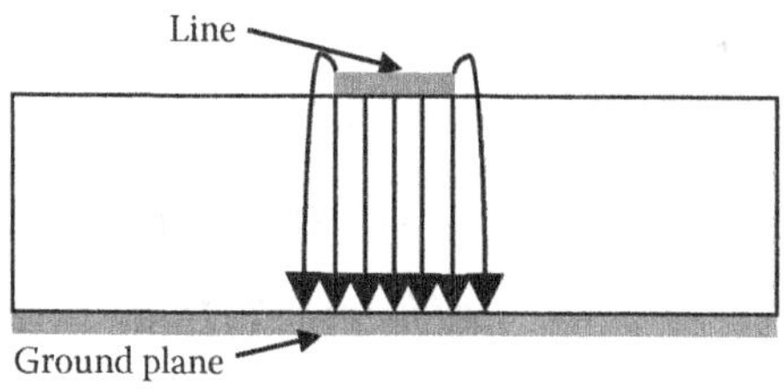

FIGURE 5.1
Electric field diagram of a microstrip transmission line.

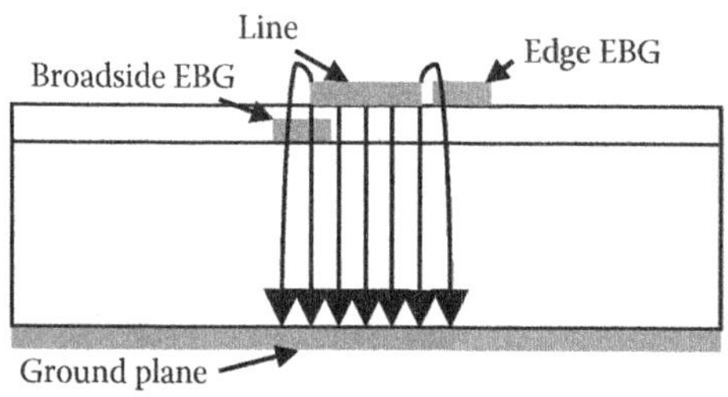

FIGURE 5.2
Electric field diagram of broadside and edge coupling of the EBG structures to a microstrip transmission line.

A cross-sectional view of the broadside-coupled EBG BSF is shown in Figure 5.3. In this multilayer design, two substrates with relative permittivity $\varepsilon_r = 2.2$ are used; the bottom substrate is 0.762 mm thick with a metallic ground plane, and the top substrate is 0.127 mm thick. The 50 Ω transmission line has a thickness of 2.76 mm on the top, and the EBG cells are located in the middle layer. The EBG cell is placed such that the overlap of the line over the cell is 0.885 mm.

When a single EBG cell (shown in Figure 2.1) is broadside coupled to a transmission line, a bandstop response is obtained. If the number of cells increases longitudinally (that is, along the transmission line), the depth of the stopband increases with no change in the stopband frequency [92,94]. The broadside-coupled EBG BSF with three EBG cells placed longitudinally is shown in Figure 5.4, and the simulated results for different numbers of cells in the longitudinal direction is shown in Figure 5.5.

Figure 5.6 shows the broadside-coupled EBG BSF with three EBG cells transverse to the line. In this case, only one EBG cell is coupled to the line while the rest are connected to the coupled cell. Figure 5.7 shows the simulated results of the broadside-coupled EBG

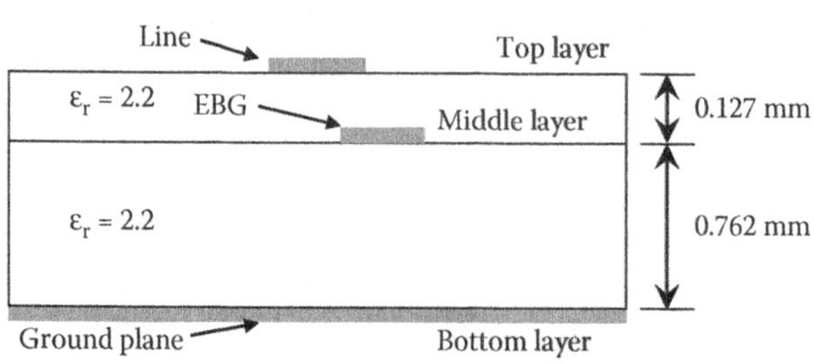

FIGURE 5.3
Cross-sectional view of multilayer broadside-coupled EBG BSF.

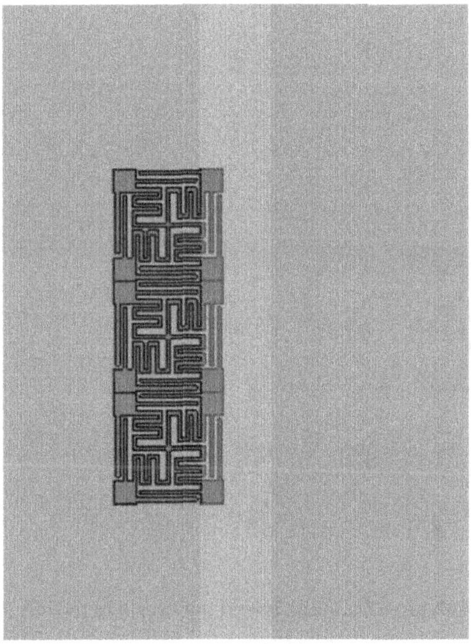

FIGURE 5.4
Broadside-coupled EBG BSF with three EBG cells placed longitudinally.

BSF with variations in the number of cells in the transverse direction. It can be seen from the results in Figure 5.7 that with a single cell only one stopband is achieved. With three cells, three stopbands are observed, but only two of them are prominent at 2.83 GHz and 4.944 GHz. With five cells, four prominent stopbands are observed. These stopbands are due to the resonances occurring in the resonator formed by the EBG cells. At these

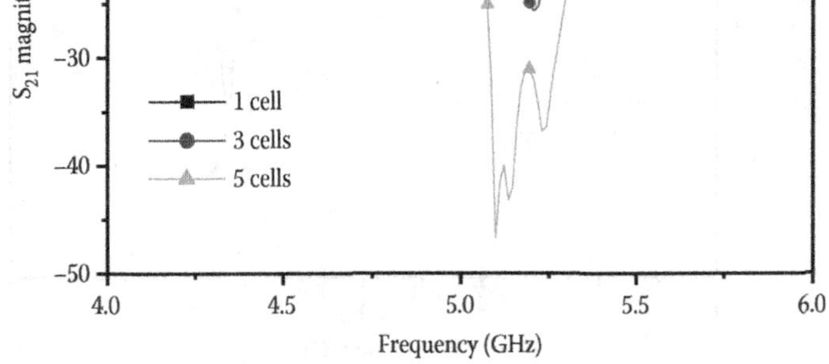

FIGURE 5.5
Simulated results of the broadside-coupled EBG BSF with variations in the number of cells in the longitudinal direction.

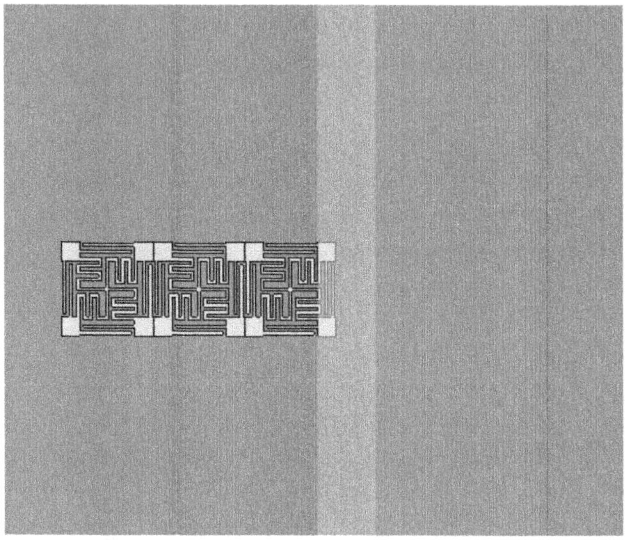

FIGURE 5.6
Broadside-coupled EBG BSF with three EBG cells placed transverse to the line.

resonance frequencies, the electromagnetic energy is confined within the resonators. This confined energy can be coupled to another transmission line to obtain BPF-like characteristics. This idea is implemented in the next section.

Figure 5.8 shows the surface current plots at stopband frequencies of a broadside-coupled EBG BSF with three EBG cells in the transverse direction. In Figure 5.8a, the current in each unit cell flows in the same direction at 2.83 GHz, which means three cells contribute to the resonance. In Figure 5.8b, at 4.94 GHz the current flows in the same direction in two

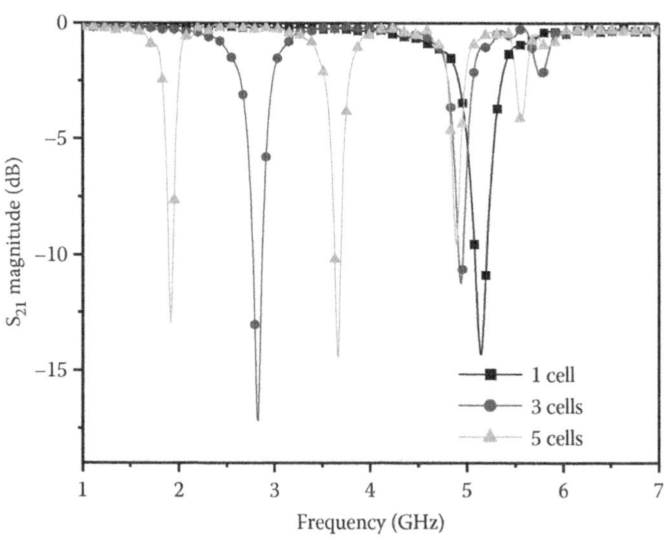

FIGURE 5.7
Simulated results of the broadside-coupled EBG BSF with variations in the number of cells in the transverse direction.

cells, while the third cell has an oppositely directed current. This shows that two unit cells are contributing to the resonance. In Figure 5.8c, at 5.76 GHz the current changes direction with each cell, which shows that each cell is contributing to the resonance.

Figure 5.9 shows a comparison of edge-coupled and broadside-coupled EBG BSFs. The same unit cell is used in both cases. In the edge-coupled arrangement, seven unit cells are

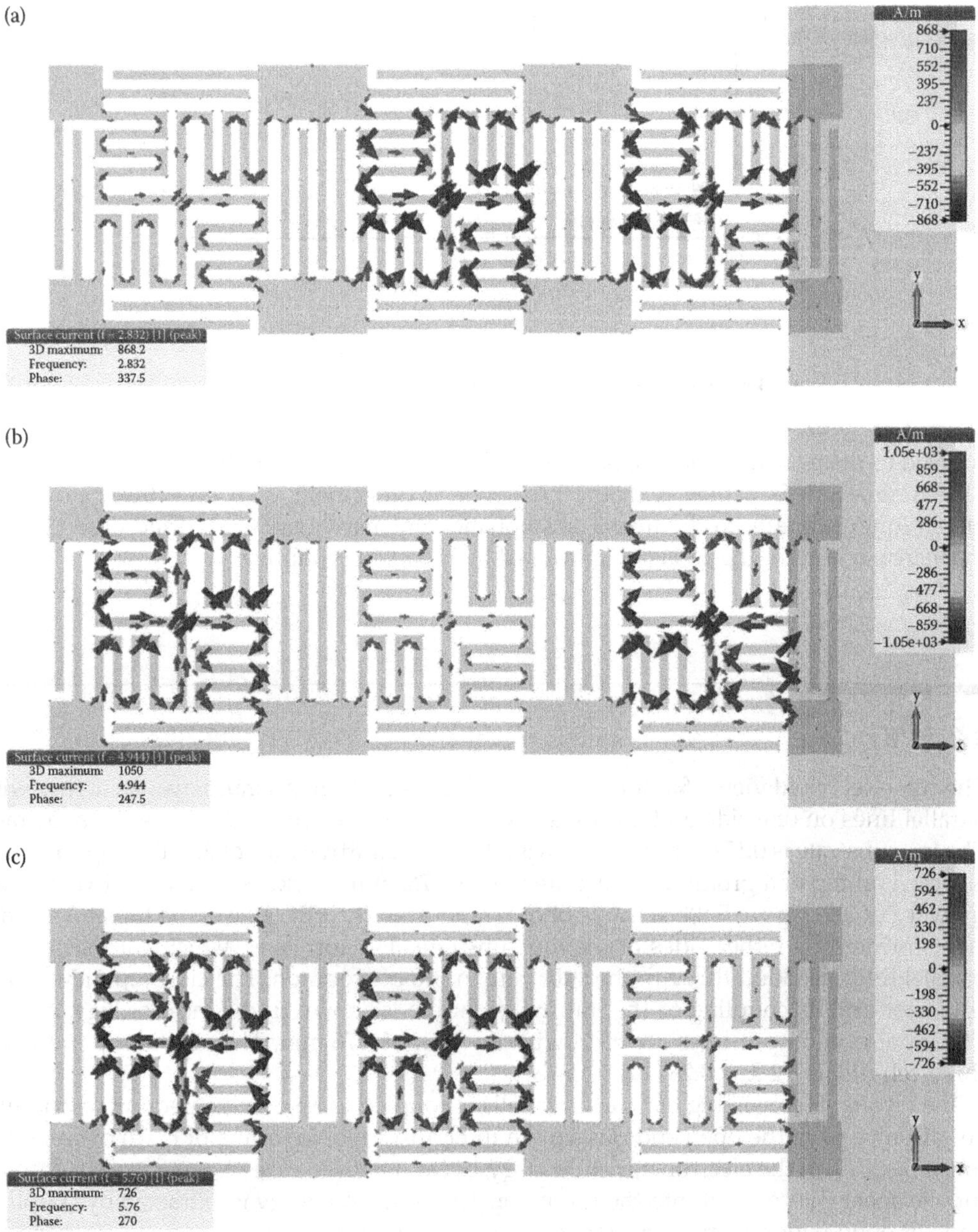

FIGURE 5.8
Surface current plots of a broadside-coupled EBG BSF with three EBG cells in the transverse direction at (a) 2.832 GHz, (b) 4.94 GHz, and (c) 5.76 GHz.

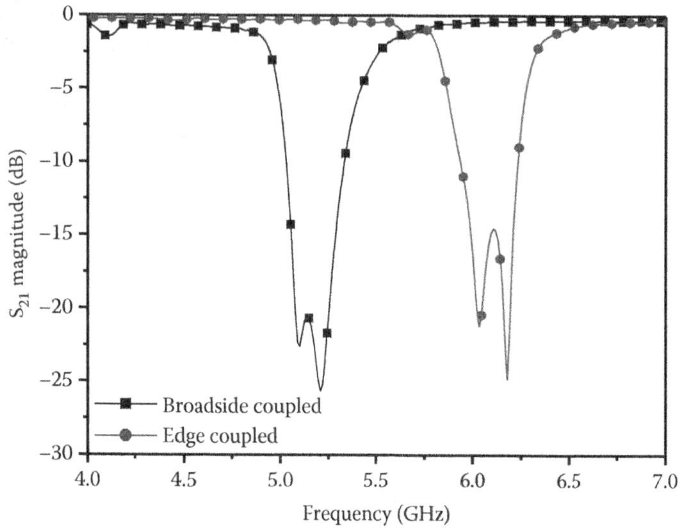

FIGURE 5.9
Comparison of edge-coupled and broadside-coupled EBG BSF.

coupled to the transmission line on either side, and a stopband depth of 20 dB is achieved. The same depth is achieved in the broadside arrangement with only three unit cells placed longitudinally to the line. The stopband center frequency is shifted down due to the increase in coupling capacitance caused by broadside coupling.

5.3 BPF

The cross-sectional view of a BPF is shown in Figure 5.10a. It is a multilayer structure with parallel lines on one side and a resonant structure on the other side of the top substrate. The top substrate is 0.127 mm thick with a relative permittivity (ε_r) of 2.2. The top substrate is placed on top of a grounded substrate with a 0.762 mm thickness and a relative permittivity (ε_r) of 2.2. Figure 5.10b shows a photograph from the EBG side of the top and middle layers printed on either sides of the top substrate. The top layer, which consists of two parallel lines, is faded through the substrate in the photograph. The close-up view of the resonator and its coupling to the transmission line is shown in the inset of Figure 5.10b. The input and output lines are 50 Ω with a width of 2.76 mm. The size of the proposed three-cell BPF is $0.2\ \lambda_g \times 0.26\ \lambda_g$.

The resonator consisting of three unit cells is coupled to an open-ended microstrip line at J distance from the open ends (as shown in Figure 5.10b) to form a BPF. An open-ended line reflects all of the electromagnetic energy, but at the resonance frequency of the resonator, the energy is coupled into the resonator. The coupled energy is extracted by coupling another open-ended microstrip line to the resonator, thus giving a bandpass response at the resonance frequency of the resonator. Because the output line is also coupled to the unit cells in a BPF, the bands are shifted slightly downward to 2.56 and 4.62 GHz unlike in the BSF with three cells (Figure 5.7).

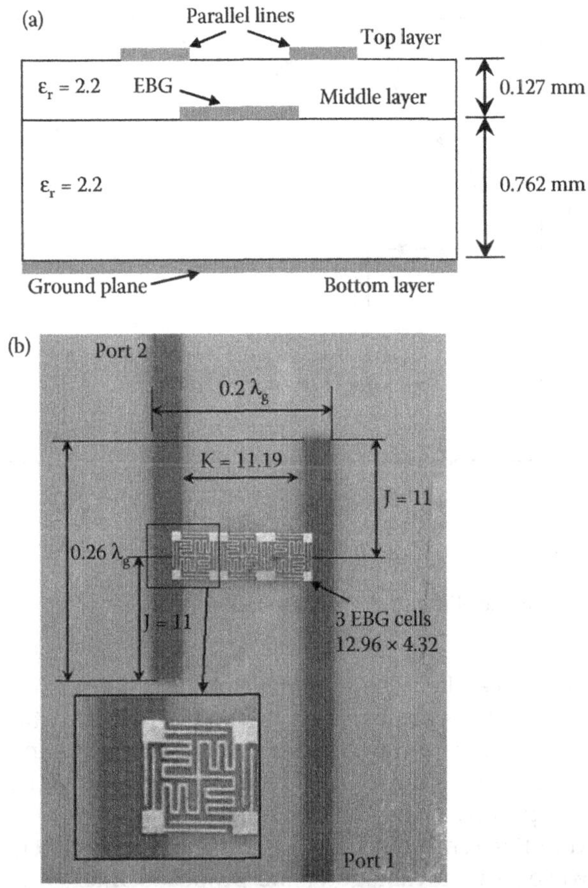

FIGURE 5.10
(a) Cross-sectional view of the BPF and (b) photograph of fabricated top and middle layers of the BPF, printed on either sides of 0.127-mm thick substrate. (All dimensions are in mm.)

Figure 5.11 shows the simulated results of the BPF with variations in extended length J. As the length J increases, the transmission at the second band (4.62 GHz) is suppressed. With J = 11 mm, the transmission reaches a minimum at 4.62 GHz. Figure 5.12 shows the simulated S_{11} of the BPF with J = 11 mm on a Smith chart and ports de-embedded up to the edge of the resonator. At 2.56 GHz (first band) the impedance is (41.19–5.24j) Ω, which is close to the center of the Smith chart, and at 4.62 GHz (second band) the impedance is (1.93+10.62j) Ω, which is close to the short circuit, as the length J (11 mm) is about a quarter wavelength at the second resonance frequency. Therefore, at the first band, the energy is coupled effectively into the resonator, and at the second band the energy is reflected back, thus forming a BPF centered around 2.56 GHz without another passband up to 11 GHz.

Figure 5.13 shows the measured and simulated results of the BPF. The measured resonant frequency is 2.56 GHz with a half-power bandwidth of 0.38 GHz and matches well with the simulations. The measured passband insertion loss is 1.5 dB, which is 0.5 dB more than the simulated values due to connector losses. The measured results indicate that harmonics are below 20 dB for frequencies up to 11 GHz. For comparison, a first-order conventional parallel coupled line BPF [123] is simulated with the same substrate parameters and is used for reference. The proposed filter has better selectivity and harmonic rejection

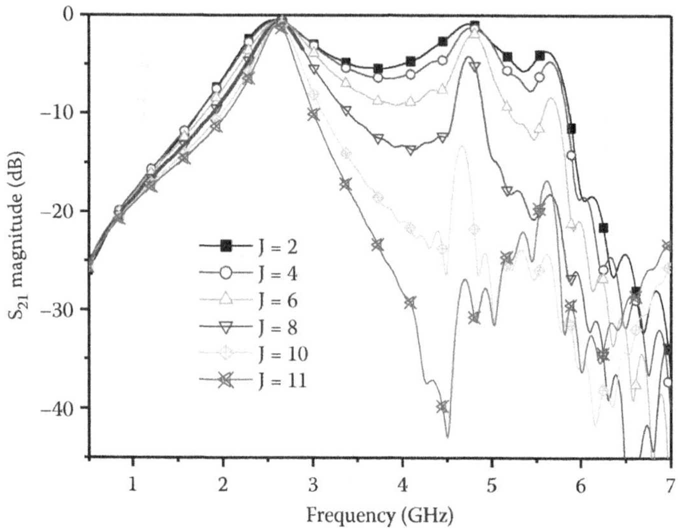

FIGURE 5.11
Transmission characteristics of the BPF with variations in J values.

capability than the reference filter. The advantage of the proposed filter lies in the fact that it does not require very small gaps between the lines as a conventional parallel coupled line BPF does, which makes fabrication easier. Moreover, to obtain the same amount of selectivity with the conventional filter, more sections are required. Hence, compactness is also achieved with the proposed coupling technique.

The filter reported in Reference 119 is similar to the proposed filter, and they are almost the same size. However, the advantage of the proposed filter as compared to the one

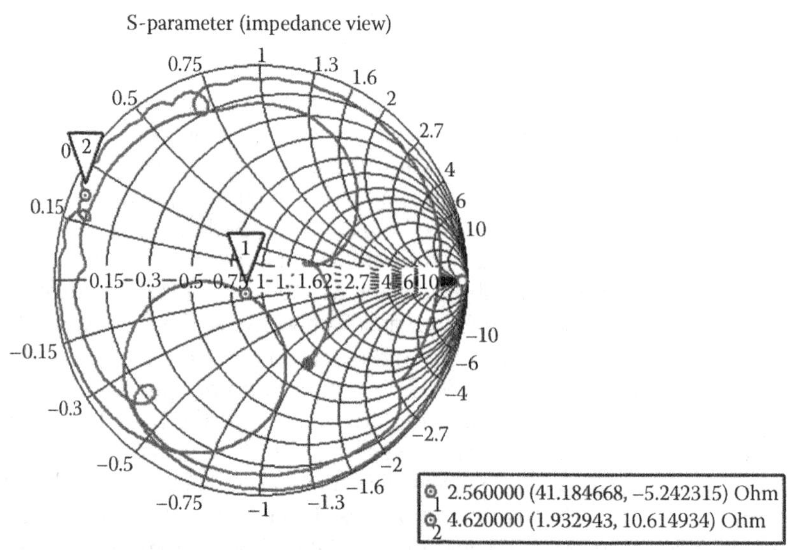

FIGURE 5.12
Simulated S_{11} of the BPF with J = 11 mm.

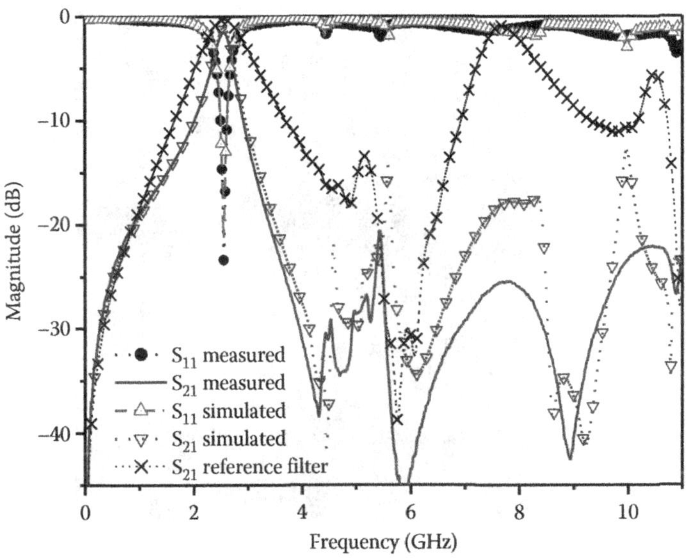

FIGURE 5.13
Measured and simulated results of the proposed BPF and the reference filter.

reported in Reference 119 is the absence of spurious passbands up to four times the resonant frequency.

The frequency of operation can be changed by changing the meander lines and capacitors in the unit cell. By increasing the number of unit cells between the parallel lines without changing the basic unit cell, the frequency of operation can be decreased. However, the lateral dimension of the filter increases. The cells are arranged in the form of a "z" to make the filter compact. Figure 5.14a shows a filter with five cells in between the parallel lines. The extended length J in the five-cell BPF is 13.5 mm. The measured and simulated results of the proposed BPF are shown in Figure 5.14b. The measured passband center frequency is 2.08 GHz with a 230 MHz bandwidth and harmonics below 10 dB for frequencies up to 11 GHz. The resonant frequency of a five-cell BPF is lower than that of the three-cell BPF, making the filter compact at a size of $0.16\lambda_g \times 0.18\lambda_g$.

5.4 Diplexer

The BPF with a wide upper stopband from the previous section is used in designing a diplexer to split or mix two signals of different frequencies. Two BPFs with resonators made up of unit cells of different sizes are used to couple different frequency bands to different output ports. A photograph of the diplexer is shown in Figure 5.15. One of the output lines connected to port 2 is coupled to the input line (port 1) by three unit cells that are each 4.32 mm × 4.32 mm in size, whereas the other output line connected to port 3 is coupled by three unit cells that are each 3.62 mm × 3.62 mm in size. The diplexer allows a 2.5 GHz signal to be diverted into port 2 and a 3.7 GHz signal into port 3.

Output port isolation in the diplexer can be improved by introducing a stopband circuit at the isolation frequency (frequency of the other output port) on the output lines. Again,

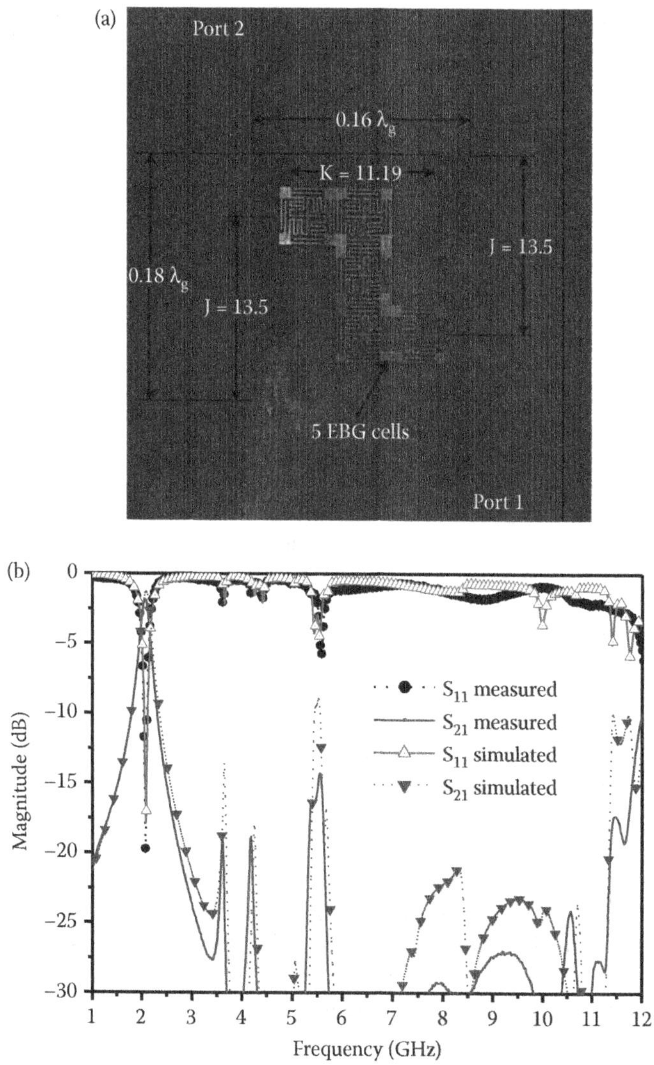

FIGURE 5.14
BPF with five cells: (a) photograph of top and middle layers and (b) measured and simulated results.

the stopband circuit is a row of three unit cells in the middle layer transverse to the line. One row of three unit cells (each 3.9 mm × 3.9 mm in size) is coupled to the line leading to port 2 to improve isolation at 3.7 GHz, and two rows of three unit cells (4.57 mm × 4.57 mm each) are coupled to the line leading to port 3 to improve isolation at 2.5 GHz. Two rows are used in the second stopband circuit to improve the depth of the stopband, as the number of cells along the line determines the depth of the stopband. The size of the unit cells used in the stopband circuit is slightly different than those used for coupling. In a stopband circuit, only one microstrip line is broadside coupled to the unit cells, whereas in the coupling circuit, both input and output lines are coupled to the unit cells. Therefore, coupling capacitance in the stopband circuit is less than the coupling capacitance in the coupling circuit, and thus slightly larger unit cells are required in the stopband circuit

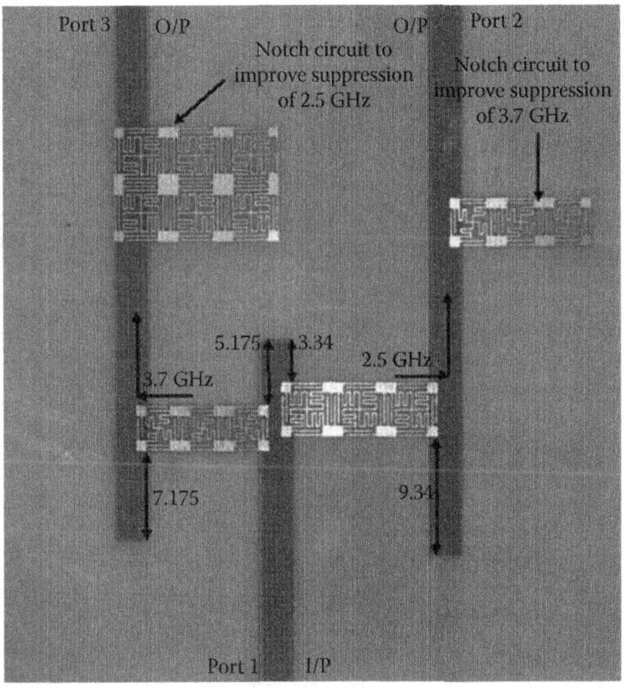

FIGURE 5.15
Photograph of fabricated top layer and middle layer of the diplexer, printed on either sides of the 0.127-mm substrate. (All dimensions are in mm.)

than in the corresponding coupling circuit for the same frequency. Figure 5.16 shows the surface current in the diplexer at 2.5 GHz and 3.7 GHz. At 2.5 GHz, more surface current is seen in the output line connected to port 2 through the resonator with 4.32 mm × 4.32 mm unit cells. Less current is seen in the other output line, and the stopband circuit with 4.57 mm × 4.57 mm unit cells further prevents the signal from reaching port 3. Thus, isolation of the 2.5 GHz signal in port 3 is improved. Similarly, at 3.7 GHz, the surface current is seen in the other channel to port 3, and the stopband circuit in port 2 helps improve the isolation of the 3.7 GHz signal into port 2.

Figure 5.17 shows the measured and simulated results of the diplexer. As observed, a 2.5 GHz signal is directed to port 2 with a measured insertion loss of 1.35 dB and a 0.45 GHz bandwidth. On the other hand, a 3.7 GHz signal is directed to port 3 with a measured insertion loss of 1.73 dB and a 0.44 GHz bandwidth. The measured isolation is greater than 18 dB in the first passband, with a 38 dB maximum isolation at 2.5 GHz. The measured isolation in the second passband is greater than 26 dB, with a 45 dB maximum isolation at 3.7 GHz. The harmonics are suppressed well below 10 dB for frequencies up to 10 GHz. If the stopband circuits are not used, the observed isolation of the 3.7 GHz signal in port 2 is 19 dB, and the observed isolation of the 2.5 GHz signal in port 3 is 15 dB. This clearly shows that the only purpose of the stopband circuit is to improve the isolation characteristics. Compared to the simulation results, the measured frequency is shifted slightly downward. The difference between the measured and simulated insertion losses can be attributed to the connector losses. From the reflection curves plotted in Figure 5.17, it is observed that above 4 GHz $|S_{11}|$ is approximately –7 dB, however at these frequencies

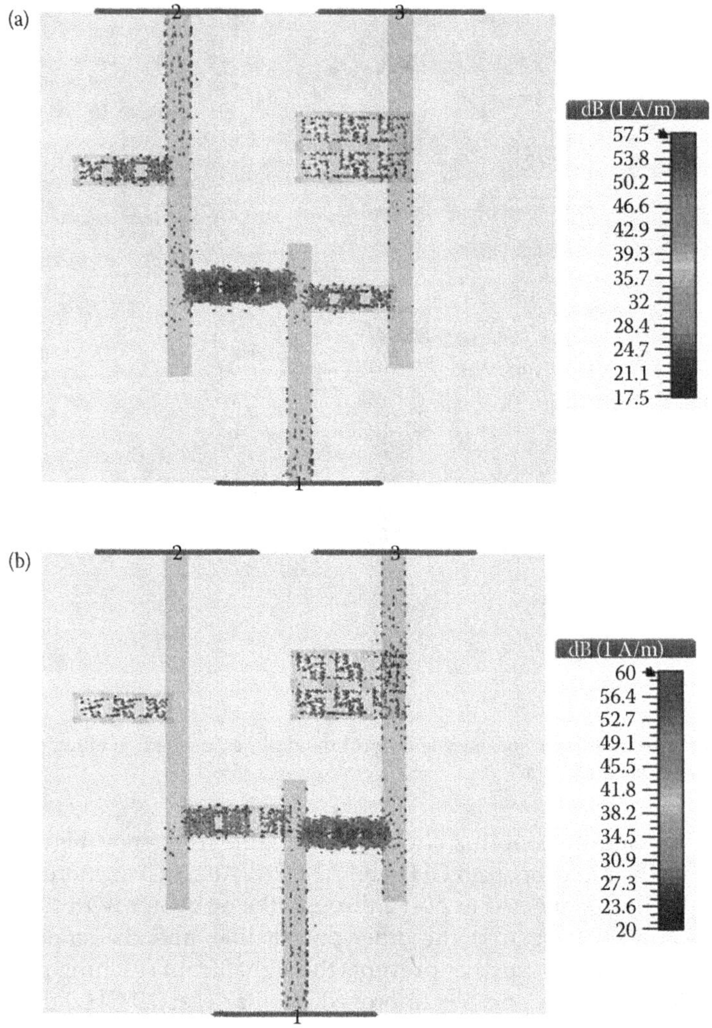

FIGURE 5.16
Surface current in diplexer at (a) 2.5 GHz and (b) 3.7 GHz.

$|S_{21}|$ is better than –10 dB. The remaining power is lost in radiation as the unit cells are made of thin lines.

The operational bands in the diplexer can be 4*f_L GHz apart, and the closeness of the operational bands are limited by the bandwidth of the passbands.

The diplexer is also tested using a spectrum analyzer, as shown in the setup in Figure 5.18. The diplexer's ports 2 and 3 are fed with signals of 2.5 and 3.7 GHz, respectively, with 0 dBm power. The two signals are combined into port 1, and observations from the spectrum analyzer show –2.00 and –2.17 dBm power at 2.5 and 3.7 GHz, respectively. The insertion loss of the radio frequency cable connected to port 1 is 0.33 and 0.5 dBm at 2.5 and 3.7 GHz, respectively. Isolation is verified by applying a 2.5 GHz signal at port 3 and a 3.7 GHz signal at port 2, and –38 and –43 dBm powers are observed in port 1 at 2.5 and 3.7 GHz, respectively. The insertion loss and isolation characteristics of the diplexer observed on the spectrum analyzer are shown in Figure 5.19.

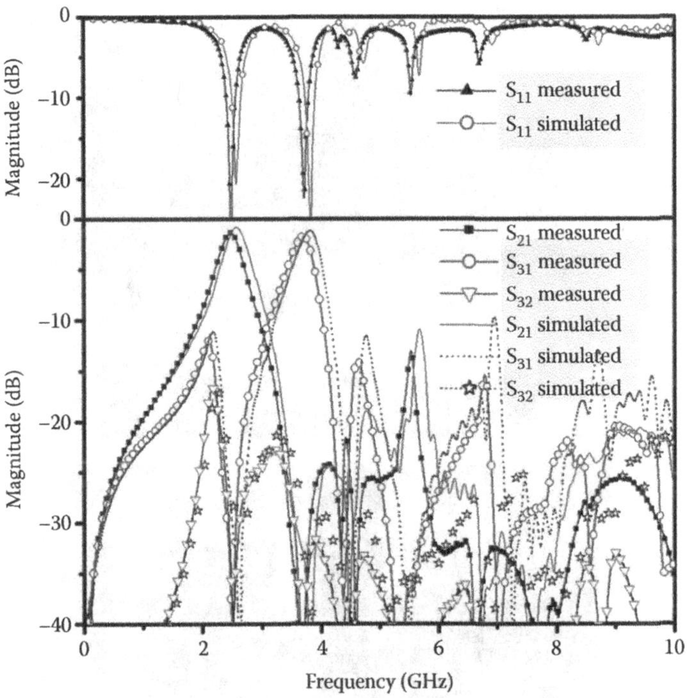

FIGURE 5.17
Measured and simulated results of the diplexer.

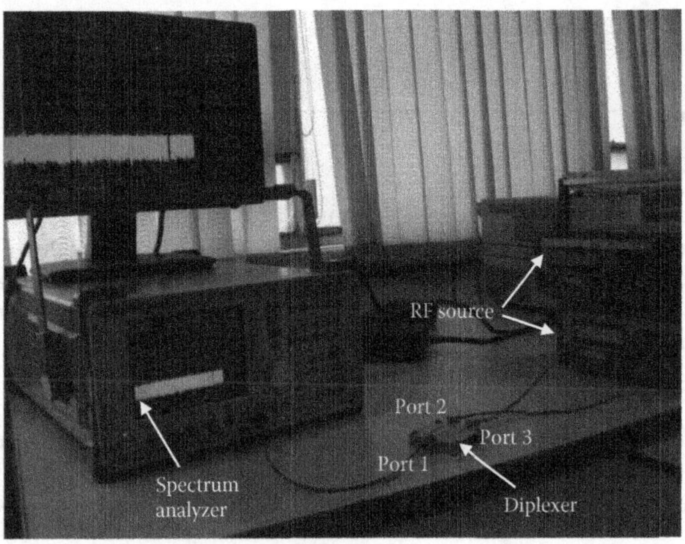

FIGURE 5.18
Diplexer measurement setup.

(a)

(b)

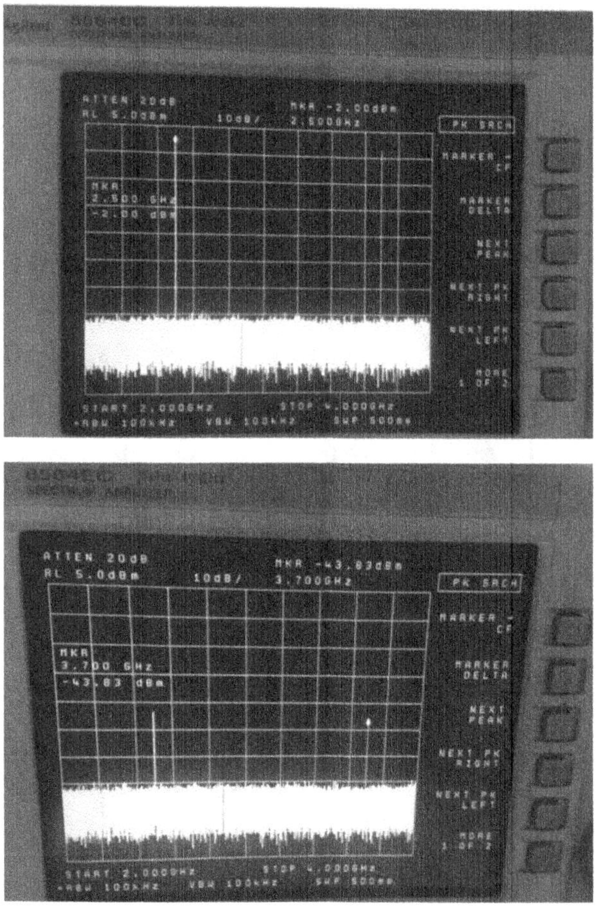

FIGURE 5.19
Measured results of the diplexer on spectrum analyzer: (a) insertion loss and (b) isolation.

5.5 Conclusion

In this chapter, EBG cells are broadside coupled to a transmission line in a multilayer design. The comparison between the broadside-coupled EBG structure and the edge-coupled EBG structure is discussed. A BSF is developed using these broadside-coupled EBG structures. A harmonic-suppressed BPF is formed by a multilayer design of the broadside-coupled EBG cells. Using the proposed BPF, a diplexer is also formed in this chapter.

Application of the EBG unit cell in the development of filters and diplexers are discussed in this chapter and previous chapters. In Chapter 6, the proposed EBG unit cell is used in an antenna application. It is shown that the directivity of the antenna is improved by using the proposed EBG cell as a frequency-selective surface screen as the superstrate to an antenna.

6

Frequency Selective Surface Application

6.1 Introduction

It is well-known that the directivity of microstrip antennas is inherently low, but it can be enhanced by using a frequency selective surface (FSS). There are two basic types of FSS structures: reflection and transmission. Reflective FSS structures are used to improve the directivity as shown in References 64–67. In References 68 and 124, the FSS exhibits transmission characteristics, and in Reference 68 the FSS is used for antenna directivity improvement.

In this chapter, FSS characteristics of the proposed unit cell are studied, and the directivity improvement of a microstrip patch antenna is presented. This chapter also discusses other applications of periodic structures.

6.2 FSS Properties of the Proposed EBG Structure

The unit cell shown in Figure 6.1 is a scaled-down version of the unit cell proposed in Chapter 2. The black design in the figure is a metal pattern on top of a GML 1000 substrate ($\varepsilon_r = 3.2$ and 0.762 mm thick) without a ground plane. The size of the unit cell is 3.51 × 3.51 mm ($0.126\lambda_0 \times 0.126\lambda_0$). The FSS characteristics of the unit cell are studied using the CST Microwave Studio simulation setup shown in Figure 6.2. The simulation setup consists of two waveguide ports placed at 5 mm ($0.18\lambda_0$) from the top and the bottom of the unit cell—that is, in the direction of the z-axis. The direction of wave propagation from the ports is normal to the plane of the unit cell. Electric boundaries ($E_t = 0$) are applied at X_{min} and X_{max}. Magnetic boundaries ($H_t = 0$) are applied at Y_{min} and Y_{max}. From the simulation results shown in Figure 6.3, it can be seen that the unit cell exhibits FSS properties, passing a 10.8 GHz signal with a 3.88 GHz 3 dB bandwidth. Thus, this structure can be useful as an FSS.

The transmission characteristics of the FSS are measured by placing the FSS between two X-band waveguides that are connected to Agilent's PNA E8364C microwave network analyzer. The measured results, shown in Figure 6.4 reveal a downward shift in the transmission band when compared to the simulated results.

The selection of the transmission frequency band through the FSS depends on the geometry of the unit cell and the associated meander lines and interdigital capacitors. The effects of the meander lines and interdigital capacitors in the unit cell are studied using three different structures (structures A, B, and C). Structure A is shown in Figure 6.1, and structures B and C are shown in Figure 6.5.

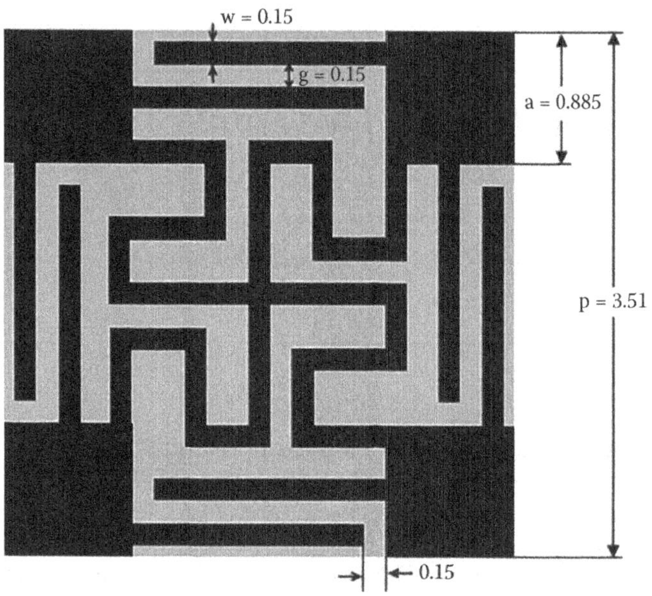

FIGURE 6.1
Unit cell (structure A). (All dimensions are in mm).

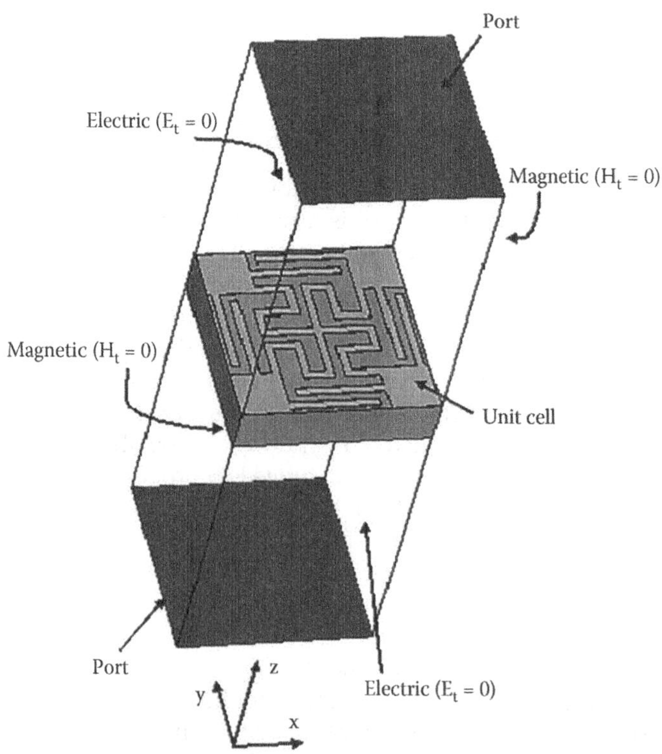

FIGURE 6.2
Simulation setup using CST Microwave Studio.

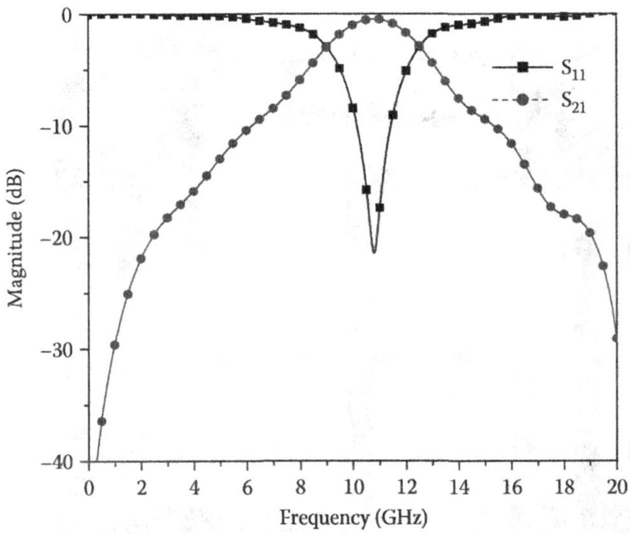

FIGURE 6.3
Simulated results of the unit cell (Figure 6.1) with the simulation setup of Figure 6.2.

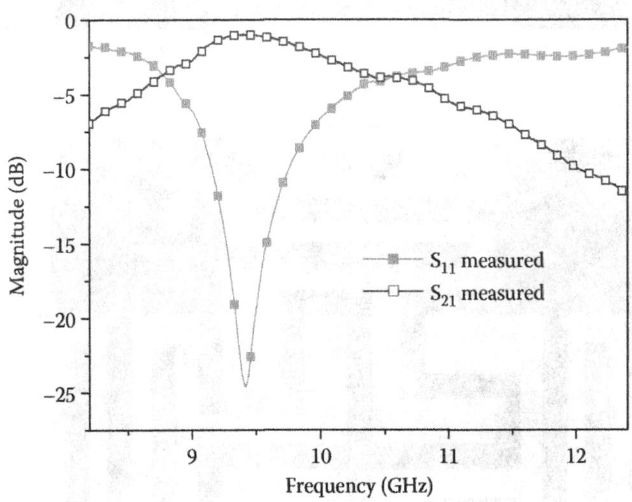

FIGURE 6.4
FSS measured results.

The overall size of the structures is the same. Figure 6.6 shows the transmission properties of all three structures simulated using the setup in Figure 6.2.

Figure 6.5a shows structure B, which is formed by replacing the meander lines from structure A with straight conductors. Keeping the capacitance of the structures the same, structure B exhibits less inductance than structure A, resulting in an upward frequency shift in the passband. Figure 6.5b shows structure C, which utilizes three-finger interdigital capacitors with a 0.1 mm gap. Since the capacitance of structure C is now greater than that of structure A and the inductance remains the same, there is a downward frequency shift in the passband. Variations in the transmission properties are similar to the bandgap variations reported in Chapter 2.

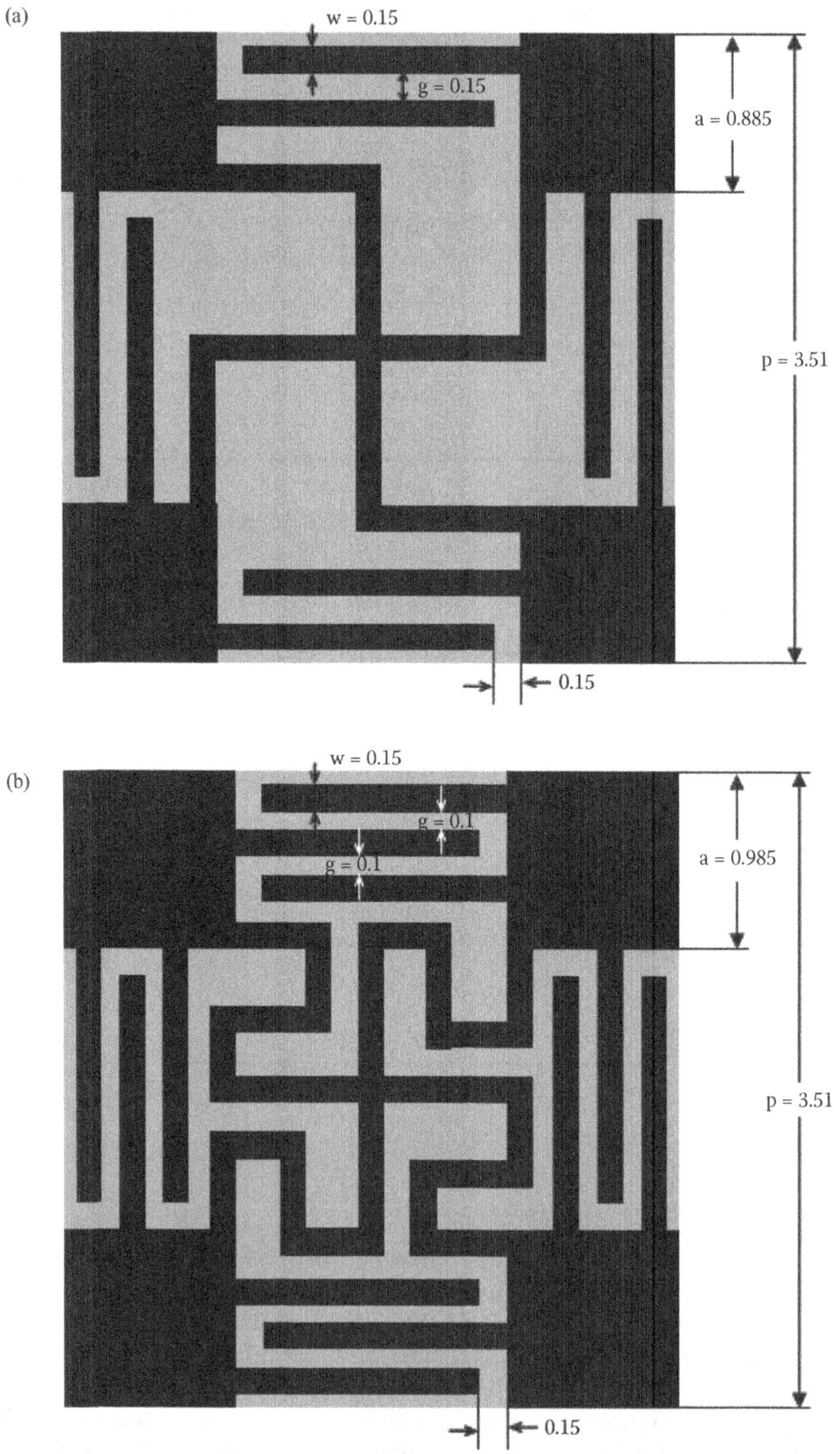

FIGURE 6.5
Unit cell. (a) Structure B and (b) structure C. (All dimensions are in mm.)

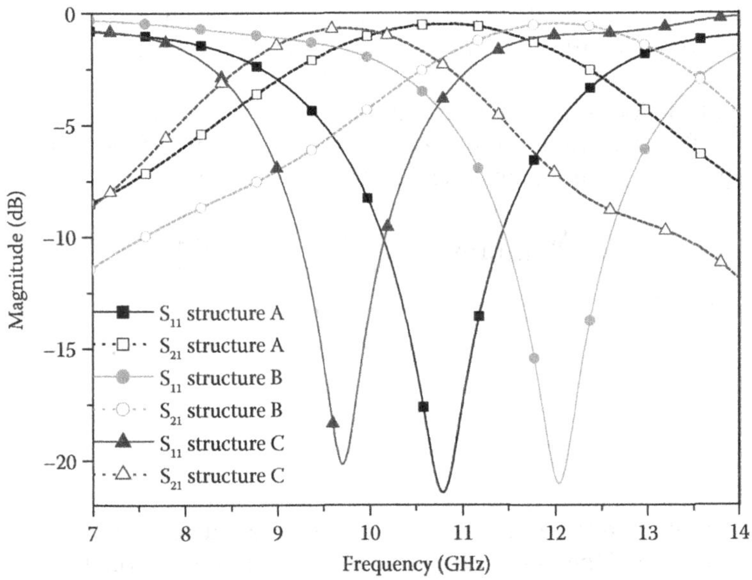

FIGURE 6.6
FSS characteristics of different unit cells.

6.3 Application of the FSS for Directivity Enhancement of the Patch Antenna

6.3.1 Inset Feed Patch Antenna

In the previous section, the FSS characteristics of the proposed structure show band-pass properties at 10.8 GHz. A conventional inset-fed patch antenna [125] is therefore designed to operate at 10.8 GHz on a GML 1000 substrate with relative permittivity $\varepsilon_r = 3.2$ and thickness $h = 0.762$ mm. The design procedure is described here.

Using a cavity model, the patch antenna is represented as a cavity with two radiating slots at the ends. The length and width of the cavity is one half guide wavelength for the dominant mode. Equation 6.1 gives an approximate value of the effective relative permittivity (ε_{re}). Equation 6.2 gives the width (W) of the patch antenna. The approximate value of ε_{re} from Equation 6.1 is used in Equation 6.2 to obtain the approximate width of the patch radiator. From there, exact ε_{re} is calculated using Equation 6.3. Multiple iterations are done using Equations 6.2 and 6.3 until W and ε_{re} are accurate to the third decimal:

$$\varepsilon_{re} \cong \frac{\varepsilon_r + 1}{2} \tag{6.1}$$

$$W = \frac{c}{2f_r\sqrt{\varepsilon_{re}}} \tag{6.2}$$

$$\varepsilon_{re} = \frac{\varepsilon_r + 1}{2} + \frac{\varepsilon_r - 1}{2}\left[1 + 12\frac{h}{W}\right]^{-1/2} \tag{6.3}$$

Because of fringing fields, the electrical length of the patch is larger than its physical length. The extended length (ΔL) for a given patch width is calculated from Equation 6.4. The effective length of the patch is calculated using Equation 6.5. Given that the resonant frequency is a function of length and that the effective patch antenna length is half the guide wavelength for the dominant mode, the physical length of the antenna (L) is calculated from Equation 6.6, taking extended length into account:

$$\frac{\Delta L}{h} = 0.412\frac{(\varepsilon_{re} + 0.3)((W/h) + 0.264)}{(\varepsilon_{re} - 0.258)((W/h) + 0.8)} \tag{6.4}$$

$$L_{eff} = L + 2\Delta L \tag{6.5}$$

$$L = \frac{c}{2f_r\sqrt{\varepsilon_{re}}} - 2\Delta L \tag{6.6}$$

The resistance at the edge of the patch (R_{in}) can be calculated from Equation 6.7, where G_1 is the conductance of the radiating slot/edge (Equations 6.8 through 6.11) and G_{12} is the mutual conductance between the two radiating slots/edges (Equation 6.12). J_0 in Equation 6.12 is the Bessel function of the first kind of zero order:

$$R_{in}(y = 0) = \frac{1}{2(G_1 \pm G_{12})} \tag{6.7}$$

$$X = k_0 W \tag{6.8}$$

$$k_0 = \frac{2\pi}{\lambda_0} \tag{6.9}$$

$$I_1 = -2 + \cos(X) + X^*S_i(X) + \frac{\sin(X)}{X} \tag{6.10}$$

$$G_1 = \frac{I_1}{120\pi^2} \tag{6.11}$$

$$G_{12} = \frac{1}{120\pi^2}\int_0^{\pi}\left[\frac{\sin((k_0W/2)\cos\theta)}{\cos\theta}\right]^2 J_0(k_0L\sin\theta)\sin^3\theta\, d\theta \tag{6.12}$$

Resistance along the length of the patch at y_0 distance from the edge is given in Equation 6.13. In order to match the 50 Ω feed line to the patch, a slot is cut in the patch up to y_0 distance from the edge, and the feed line is connected there as shown in Figure 6.7.

$$R_{in}(y = y_0) = R_{in}(y = 0)\cos^2\left(\frac{\pi y_0}{L}\right) \tag{6.13}$$

From these equations, the patch antenna dimensions are calculated to operate at 10.8 GHz. Using CST Microwave Studio, the dimensions are optimized to adjust the

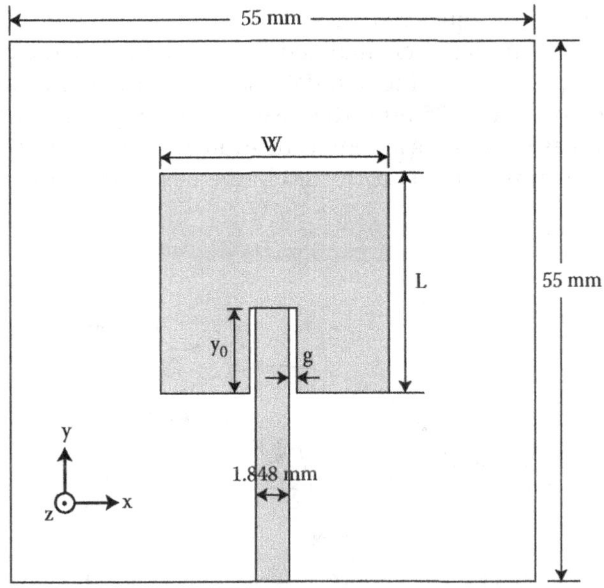

FIGURE 6.7
Schematic diagram of the patch antenna.

antenna resonant frequency to exactly 10.8 GHz. The schematic diagram of the patch antenna is shown in Figure 6.7. The dimensions of the patch antenna are W = 8.23 mm, L = 7.62 mm, and inset length (y_0) = 2.654 mm. The gap in the inset between the feed line and the patch (g) is 0.46 mm. The width of the 50 Ω feed line is 1.848 mm. The simulated and measured S_{11} of the patch antenna is shown in Figure 6.8.

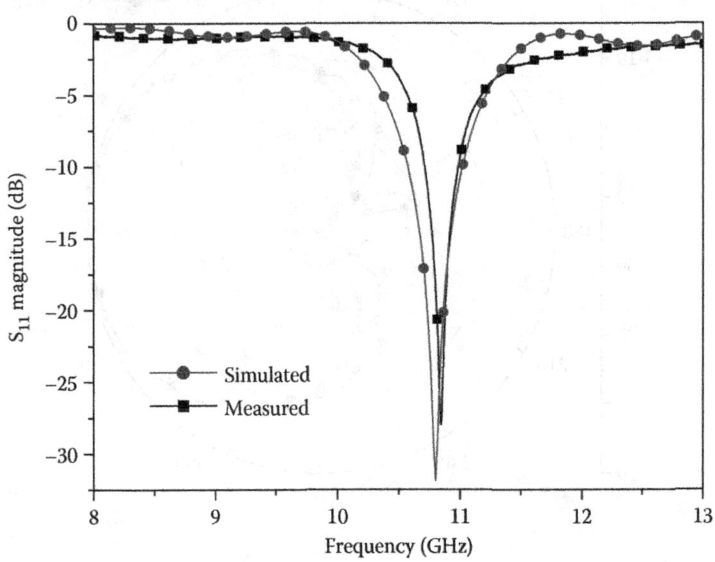

FIGURE 6.8
Simulated and measured S_{11} of the patch antenna.

The measured resonant frequency is 0.045 GHz higher than the simulated value. Both the measured and the simulated normalized E-plane and H-plane patterns are shown in Figure 6.9. It is apparent from the simulations that with an increase in the substrate and ground plane size above 35 mm, the beam in the E-plane starts splitting. In the event that the circuit as well as the antenna substrate must be large, the beam splitting is unavoidable. This beam splitting is corrected in the next section by using an FSS screen as a superstrate.

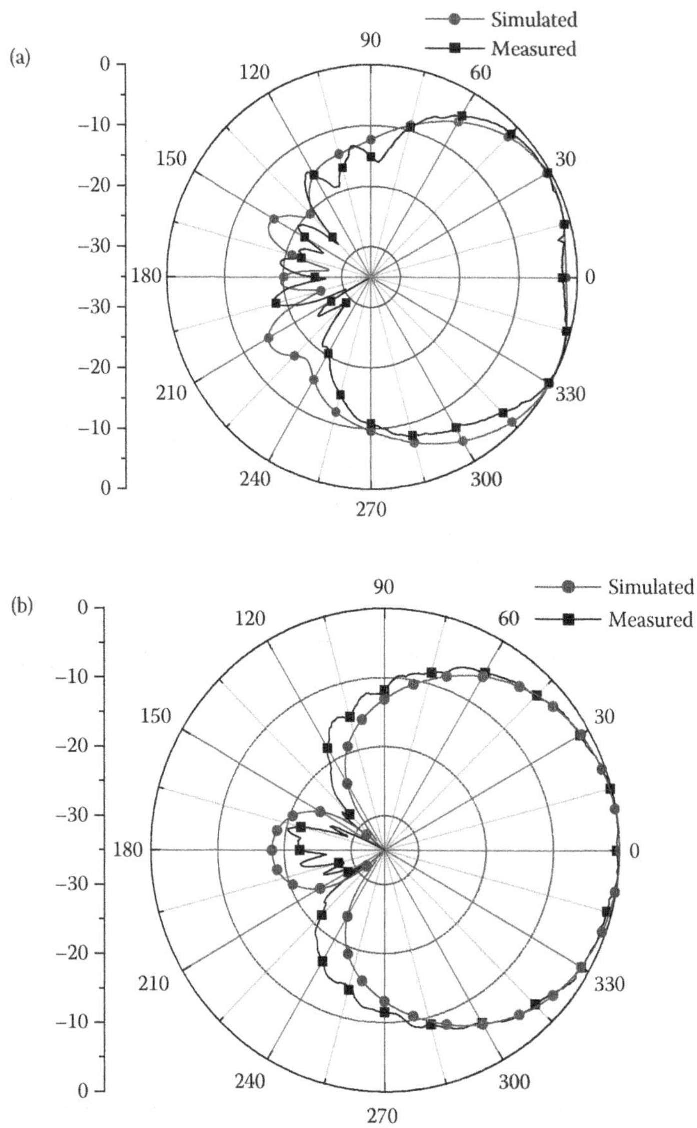

FIGURE 6.9
Simulated and measured radiation patterns of the patch antenna at 10.8 GHz: (a) E-plane (y–z plane: $\varphi = 90°$) and (b) H-plane (x–z plane: $\varphi = 0°$).

6.3.2 Patch Antenna with FSS as Superstrate

An FSS screen is fabricated with a 13 × 13 array of unit cells (structure A) on a GML 1000 substrate measuring 55 × 55 mm. The screen is used as a superstrate over the antenna at a distance d supported by polystyrene. A photograph of the patch antenna with the superstrate is shown in Figure 6.10.

The distance d is optimized for maximum directivity. The simulated radiation patterns of the patch antenna with the FSS superstrate for various values of d at 10.8 GHz are plotted in Figure 6.11. It is observed that maximum directivity occurs at d = 14.5 mm ($\approx 0.5\lambda_0$) and its integer multiples. Therefore, the superstrate is maintained at 14.5 mm above the patch antenna for further experiments.

The directivity of the patch antenna with the FSS screen as a superstrate is enhanced because the FSS screen and patch antenna are acting as a cavity resonant antenna. The FSS screen is placed over the patch antenna at a distance d, supported by spacers as shown in the photograph in Figure 6.10. The FSS screen partially reflects the signal at 10.8 GHz, as can be seen from the measured results in Figure 6.4. As the radiations from the patch antenna operating at 10.8 GHz strike the FSS, some are transmitted and the rest are reflected back. The reflected wave bounces back toward the FSS by a ground plane. When the total phase due to the distance between the FSS and the ground, the reflection phase of the FSS, and the 180° phase reflection from the ground becomes integral multiples of 2π, the waves passing through the FSS will be coherent, resulting in enhanced directivity.

The reflection coefficient of the patch antenna with and without a superstrate is measured using an Agilent PNA E8364C and is shown in Figure 6.12. The resonant frequency of the patch antenna with a superstrate is shifted slightly downward due to loading with the superstrate. The measured and simulated E-plane and H-plane patterns of the antenna

FIGURE 6.10
Patch antenna with the FSS screen as a superstrate.

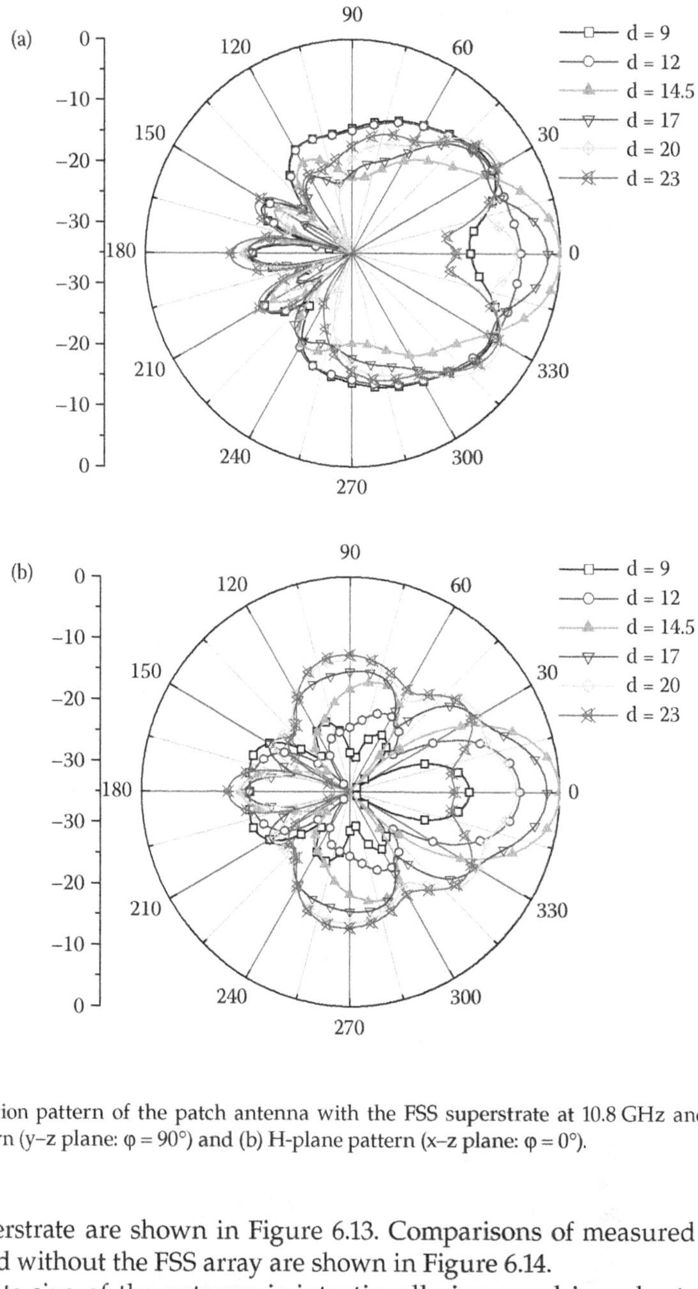

FIGURE 6.11
Simulated radiation pattern of the patch antenna with the FSS superstrate at 10.8 GHz and variations in d: (a) E-plane pattern (y–z plane: φ = 90°) and (b) H-plane pattern (x–z plane: φ = 0°).

with the superstrate are shown in Figure 6.13. Comparisons of measured radiation patterns with and without the FSS array are shown in Figure 6.14.

The substrate size of the antenna is intentionally increased in order to account for a larger ground plane size, as is required by most transceiver systems. As a result, the E-plane beam is almost flat around the broadside direction with a maximum at ±25°. With the superstrate, the maximum is brought back to normal direction (0°) in the E-plane. The simulated directivity of the patch antenna with and without the FSS is 14.39 dBi and 7.44 dBi, respectively. The improvement in the directivity is 6.95 dB.

Measured directivity is calculated numerically from the two measured principle planes patterns of the antenna without and with FSS is 5.6 dB and 13.6 dB. An improvement of 8 dB is seen in the measurement. Because of only 4 samples in azimuthal plane in measured

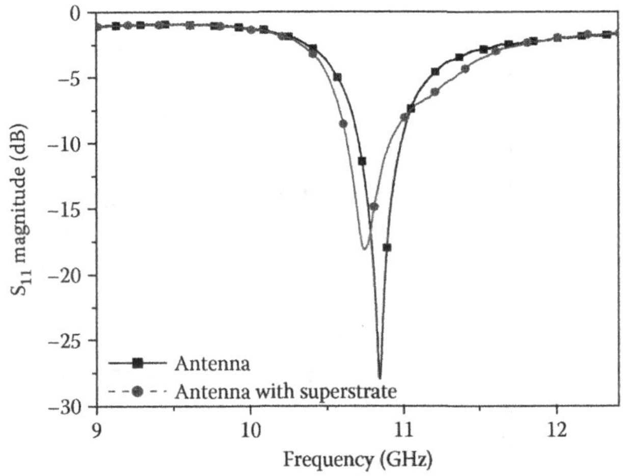

FIGURE 6.12
Reflection coefficient of the patch antenna with and without the FSS superstrate.

calculations, 1 dB difference is seen in the measured and simulated. Front-to-back ratios (FBR) of the patch antenna are 22 dB and 22.8 dB in the H-plane and the E-plane, respectively. FBRs for the patch antenna with the FSS are 19.8 dB and 16.2 dB in the H-plane and the E-plane, respectively. FBRs of the antenna with the FSS have somewhat more variation because of the slight alignment error of 3°. On the whole, the FSS has improved the directivity by 6.5 dB while sacrificing the FBR slightly.

To identify the exact role of the FSS structure, simulations are carried out with only a dielectric as a superstrate over the antenna at the distance of 14.5 mm from the antenna. Figure 6.15 shows simulated radiation patterns of the antenna, the antenna with the dielectric superstrate, and the antenna with the FSS superstrate. Patterns are normalized with respect to the maximum of antenna with FSS structure. It is seen that the antenna with dielectric superstrate shows hardly any improvement in the directivity as compared with the reference antenna without any superstrate. However, with the proposed FSS there is significant increase in power along broadside. These results clearly show the benefits of the FSS screen for directivity enhancement.

Power flow in the patch antenna is observed using CST simulation software. Figure 6.16 shows power flow in a plane perpendicular to the z-axis (that is, top view) for the reference patch antenna, the patch antenna with the dielectric, and the patch antenna with the FSS superstrate. The power flow observation plane is kept at 8 mm from the reference patch and 20 mm above the patch antenna with the superstrate. The arrows in Figure 6.16 indicate power flow from the antenna after passing through the superstrate.

A comparison of power flow in the three cases reveals that with the FSS superstrate, the waves get bent (refracted toward the normal), leading to more power in the forward direction along 0°. This results in improved directivity along the broadside.

To see the effect of the FSS screen in its stopband, the FSS screen is used as a superstrate at 25 mm distance to the patch antenna operating at 6.7 GHz. The measured radiation patterns of the 6.7 GHz patch antenna with and without the FSS superstrate are shown in Figure 6.17.

The figure shows that the FSS screen is decreasing the directivity of the antenna operating at 6.7 GHz in the main/forward direction. The FBRs of the antenna operating at

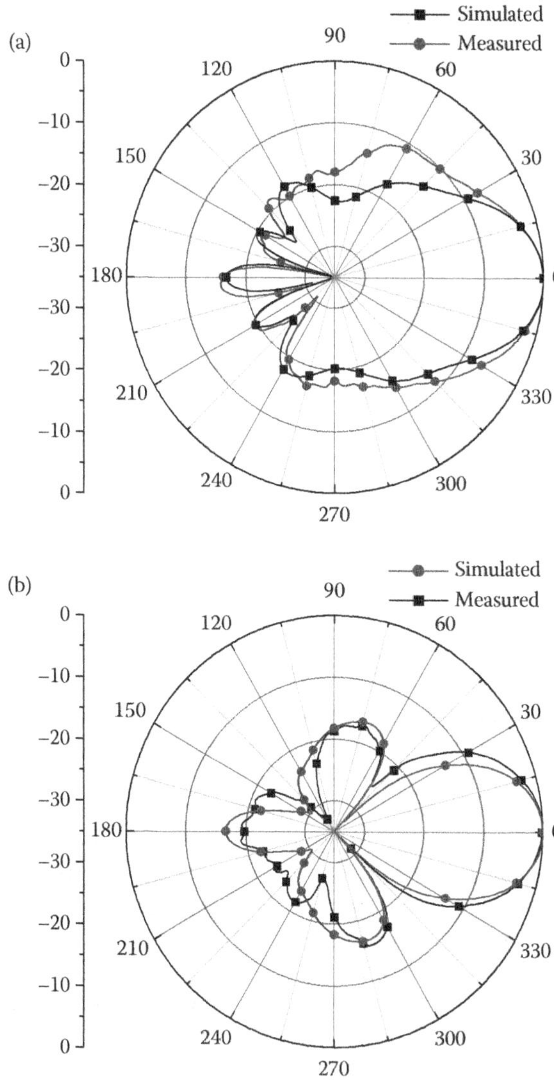

FIGURE 6.13
Measured and simulated radiation patterns of the patch antenna with the FSS superstrate at 10.8 GHz: (a) E-plane pattern (y–z plane: $\varphi = 90$) and (b) H-plane pattern (x–z plane: $\varphi = 0°$).

6.7 GHz with and without the FSS superstrate in the E-plane are −0.28 dB and 15.9 dB, respectively, and in the H-plane, the values are 3.37 dB and 18.4 dB, respectively. Thus, it is clearly understood that the FSS screen is reflecting 6.7 GHz signals.

A comparison of this work with other reported works is given in Table 6.1 [65,66,68,124]. The proposed structure in this work is compact and has higher directivity than others. The structure in Reference 124, however, is not used for antenna applications.

6.3.3 Coplanar Waveguide (CPW)-Fed Monopole with FSS as Reflector

Artificial magnetic conductors (AMC) or high impedance surfaces (HIS) are structures created by a periodic arrangement of a unit cell with unique properties in one, two, or

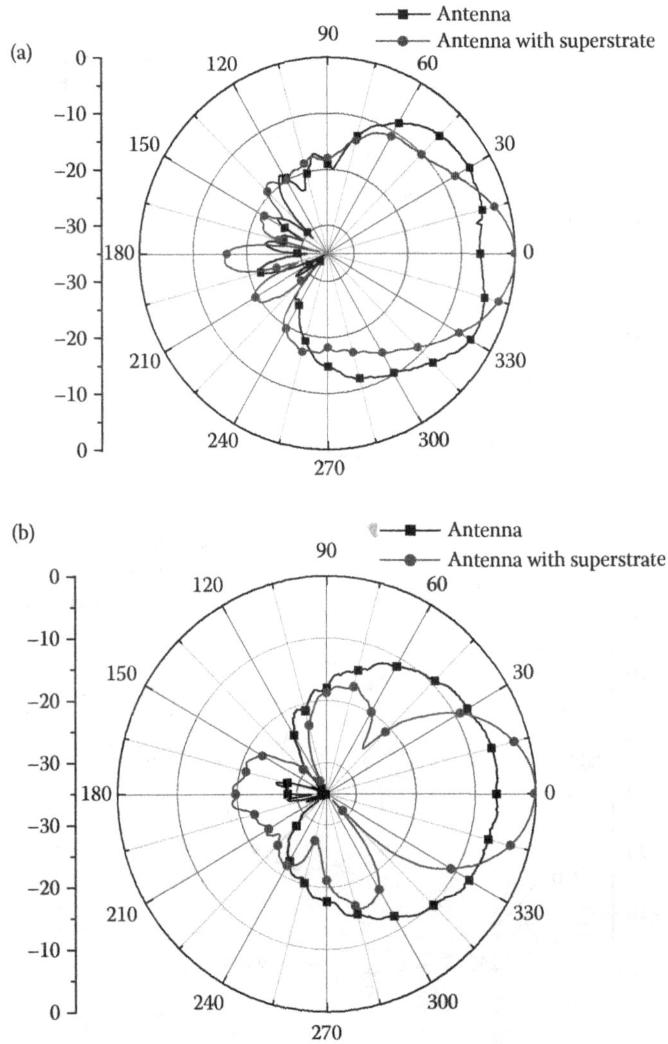

FIGURE 6.14
Measured radiation pattern of the patch antenna with and without the FSS superstrate at 10.8 GHz: (a) E-plane pattern (y–z plane: $\varphi = 90°$) and (b) H-plane pattern (x–z plane: $\varphi = 0°$).

three dimensions. The unit cell is usually made up of conducting patches. Slots of various shapes are embedded in these patches to create different shapes with varying properties. The most popular shapes among these are the square ring, the Jerusalem cross, and so on. This unit cell is repeated in two dimensions to create an HIS. The reflection and transmission properties of such surfaces are attracting a lot of attention for their potential antenna applications. One of the distinct features of these surfaces that can be employed in antenna systems is in-phase reflection with near unity amplitude in the designed band of interest.

In References 126–129 and other similar reports, the authors have demonstrated improvement in antenna impedance matching using AMCs/HISs. The very popular planar inverted-F antenna (PIFA) reportedly exhibited improved performance over AMCs

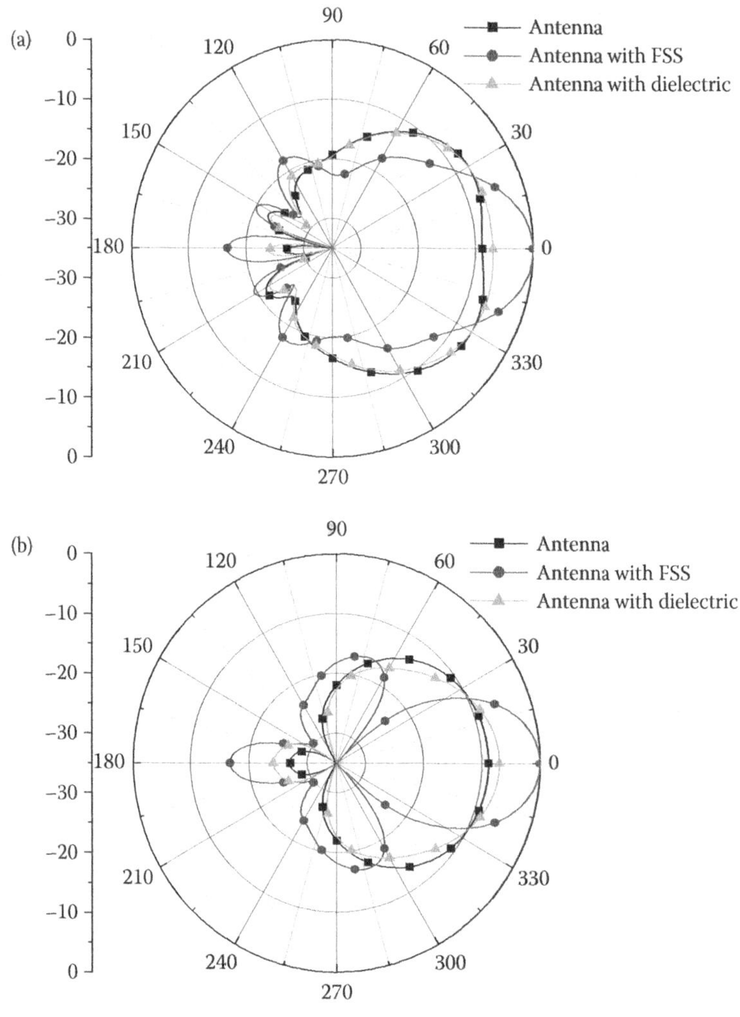

FIGURE 6.15
Simulated radiation patterns of the patch antenna, patch antenna with the dielectric superstrate, and patch antenna with the FSS superstrate operating at 10.8 GHz: (a) E-plane pattern (y–z plane: $\varphi = 90°$) and (b) H-plane pattern (x–z plane: $\varphi = 0°$).

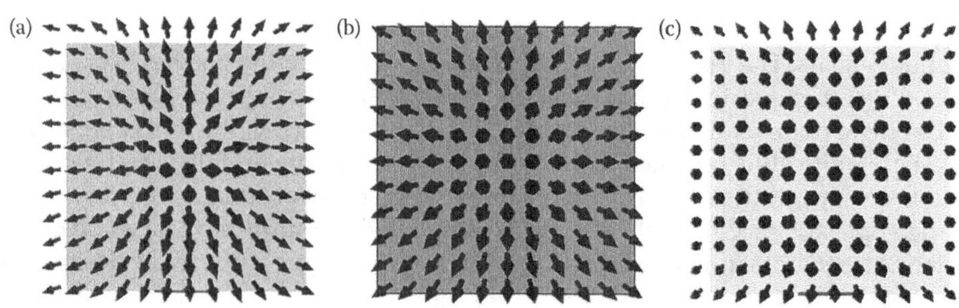

FIGURE 6.16
Power flow top view of (a) patch antenna, (b) patch antenna with dielectric superstrate, and (c) patch antenna with FSS superstrate.

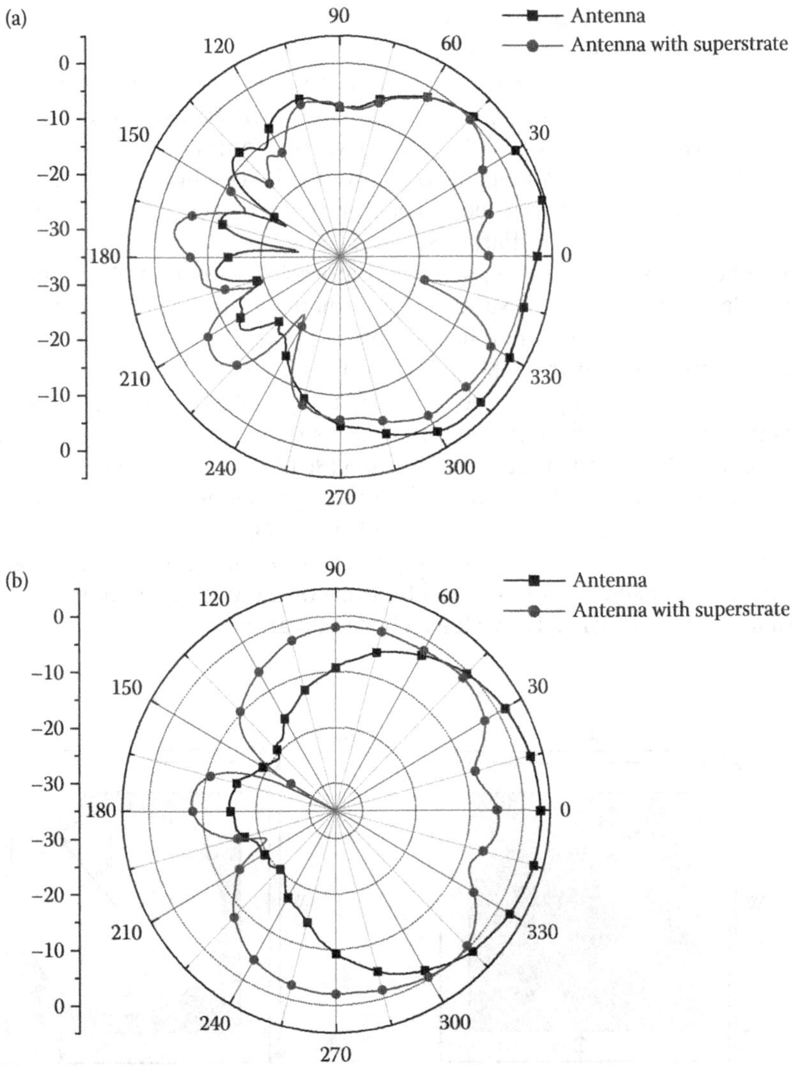

FIGURE 6.17
Measured radiation pattern of the patch antenna with and without the FSS superstrate operating at 6.7 GHz: (a) E-plane pattern (y–z plane: $\varphi = 90°$) and (b) H-plane pattern (x–z plane: $\varphi = 0°$).

TABLE 6.1
Comparison of FSSs Showing Improvement in the Antenna Directivity

Work	Size of Unit Cell	Substrate (ε_r, Height, mm)	Distance of FSS from Antenna (mm)	Directivity Improvement (dBi)	FSS Type
This work	$0.126\lambda_0$	3.2, 0.762	$0.52\lambda_0$	7	Transmission
[65]	$0.224\lambda_0$	3.38, 0.782	$0.47\lambda_0$	6	Reflective
[66]	$0.12\lambda_0$	4.4, 1	$0.26\lambda_0$	2.5	Reflective
[68]	$0.135\lambda_0$	10.2, 1.27	$0.49\lambda_0$	3	Transmission
[124]	$0.082\lambda_0$	2.65, 1	NA	NA	Transmission

[130]. There have been numerous reported improvements in impedance and bandwidth as well as gains with planar antennas using HISs [131–135].

Recent advances in AMC/HIS structures are reported in References 136–142. However, these structures exhibit narrow in-phase reflection conditions, so the real challenge is to improve the bandwidth to make AMCs/HISs useful in applications like ultra-wideband (UWB) antennas. The other challenge is to obtain incidence-angle and polarization insensitivity. Moreover, these structures need to be thin in the direction of incidence so as not to increase the overall size of the radiating system.

Figure 6.18 shows the evolution steps of the proposed dual-band AMC structure. We start with the regular square patch (AMC-A) with dimensional variables as shown in Figure 6.18a. The other side of the substrate has a ground plane. In the next step, we introduce a circular slot as shown in Figure 6.18b (AMC-B). This increases the effective inductance of the structure, thereby shifting the ±90° reflection phase AMC band to the lower side. In the final step, a circular patch with a smaller radius than the circular slot is introduced to establish another AMC band, as seen in Figure 6.18c (AMC-C). This not only introduces a second band with a wider AMC bandwidth but also provides the possibility of tuning its center frequency. Although the proposed structure (AMC-C) is a dual-band structure, we have used only the second band (higher frequency) for application in the planar monopole antenna. There have been reports in the literature of the use of a square ring slot, but to achieve polarization insensitivity, we have chosen an annular slot.

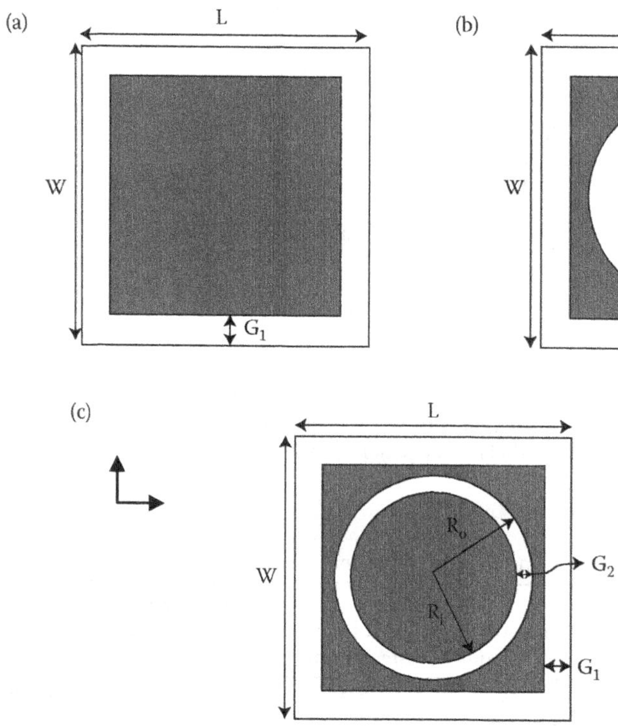

FIGURE 6.18
Evolution of the proposed structure. (a) Rectangular square patch structure: AMC-A, (b) AMC-A with a circular slot: AMC-B, and (c) AMC-B with a circular patch (or AMC-A with an annular slot): AMC-C. Dimensions (in mm): $L = W = 22$, $G_1 = 0.5$, $R_o = 10.45$, $R_i = 9.6$, and $G_2 = 0.85$.

The proposed structure was analyzed using CST Microwave Studio. A 3.048 mm thick dielectric with $\varepsilon_r = 3.2$ is used to assemble the structure. The reflection-phase characteristics of the three AMC structures are shown in Figure 6.19. The reflected phase from the AMC surface (Φ_{AMC}) is normalized to that of a perfect electric conductor (PEC) (Φ_{PEC}) structure by using the relation expressed in Equation 6.14 to cancel the propagation phase that exists due to the distance between the input and load of the waveguide. We need to add 180° to account for the reflection phase of the PEC. This procedure is reported in Reference 124, although in a different context:

$$\Phi = \Phi_{AMC} - \Phi_{PEC} + 180^\circ \tag{6.14}$$

It can be seen that AMC-A exhibits a ±90° reflection phase with the AMC band of 400 MHz bandwidth centered at 2.65 GHz. As expected, the AMC bandwidth is narrow for this structure. Upon introducing the circular slot (AMC-B), the AMC band shifts to a lower frequency band centered at 1.87 GHz. This can be considered miniaturizing the structure. However, the bandwidth also decreases to 80 MHz.

Introduction of the circular patch at the center loads the inductance of AMC-B with an LC resonator in parallel as shown in Figure 6.20. This reduces the overall capacitance of the structure, resulting in an upward shift in the AMC frequency band of AMC-B (see Figure 6.19). The AMC band of the structure also increases if only the circular patches are kept in place and all other metal portions of the structure are removed. This scenario is depicted in Figure 6.19 for reference. The resonance frequencies of the equivalent circuit in Figure 6.20 can be obtained from a quadratic equation:

$$\omega^4(L_1C_1L_2C_2) - \omega^2[L_1C_1 + C_2(L_1 + L_2)] + 1 = 0 \tag{6.15}$$

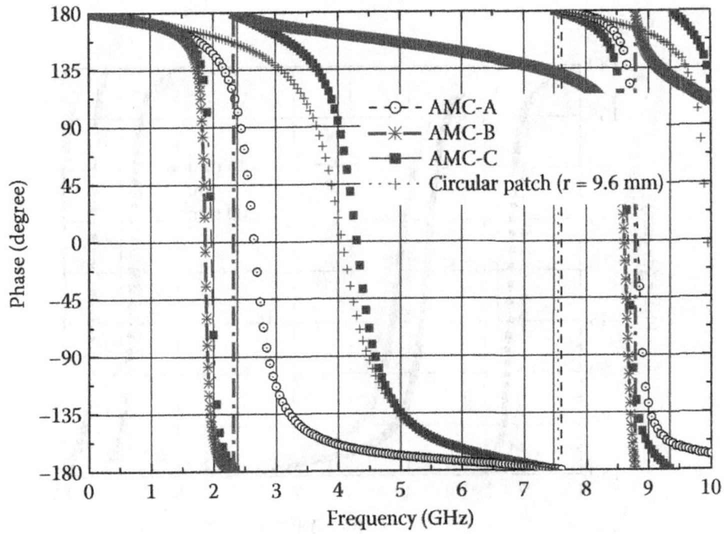

FIGURE 6.19
Simulated reflection-phase characteristics of the three AMC structures. The reflection phase of a structure consisting of only circular patches is also shown for justifying the origin of two AMC bands.

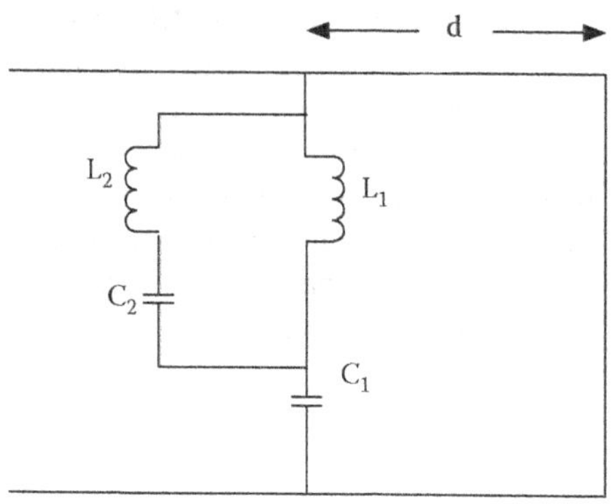

FIGURE 6.20
Equivalent circuit of the proposed AMC-C indicating additional capacitance in series, which reduces the center frequency of the AMC band.

The AMC-C structure now has two bands: that of the AMC-B structure centered at 1.97 GHz with a very narrow bandwidth and that of the circular patch centered at 4.3 GHz with a 600 MHz bandwidth. It must be mentioned here that the second band can be tuned by connecting varactors across the annular slot to vary its capacitance, as is done in Reference 143. This is verified by changing the annular slot-width G_2 and shown in Figure 6.21, where a tuning range of about 700 MHz is observed.

Polarization sensitivity studies are done only at the second AMC band of AMC-C by placing the AMC structure at the end of a C-band waveguide and then rotating it 90°. The results are presented in Figure 6.22; although there are four different curves, only two are seen because the other two are exactly identical. It must be mentioned here that the measured results are for a 2 × 1 array of the unit cell (dictated by the size of a C-band waveguide).

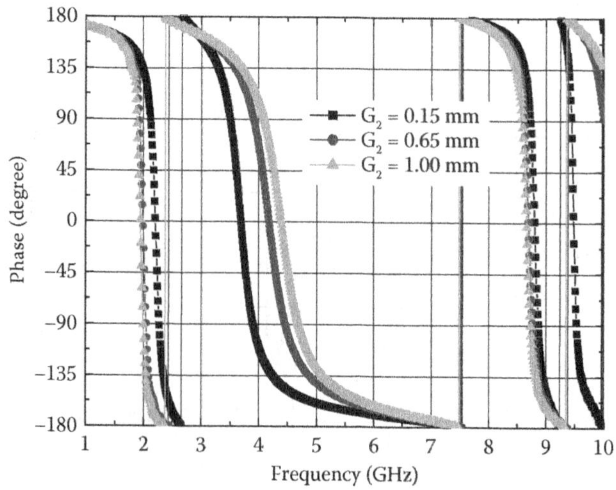

FIGURE 6.21
Effect of series capacitance due to the annular slot on the reflection-phase characteristics of the proposed structure (AMC-C).

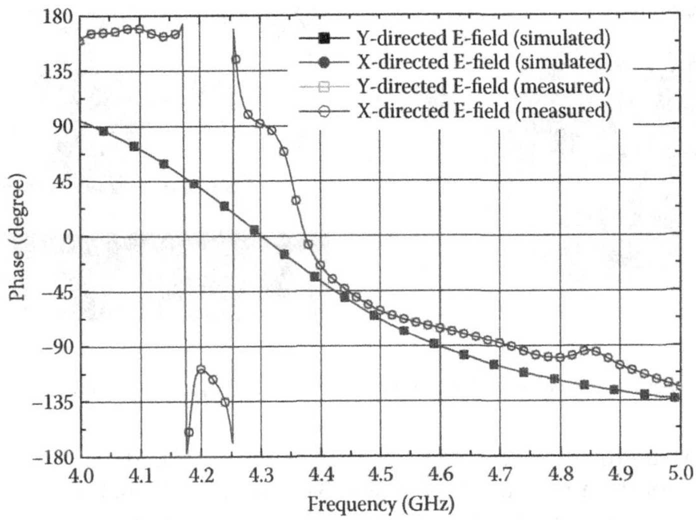

FIGURE 6.22
Effect of the polarization of the incident wave on the reflection properties of AMC-C.

Figure 6.22 shows that, when measured, the second AMC band is centered at 4.36 GHz with a bandwidth of 400 MHz. This band is used as a back-plane for a CPW-fed monopole antenna.

A standard patch monopole antenna excited by a CPW line is designed. Although, the proposed AMC structure exhibits dual-band characteristics, the bandwidth of the first AMC band is very narrow. Hence, the antenna is designed to resonate at around the center frequency of the second AMC band. The substrate for the antenna is 0.762 mm thick with relative permittivity $\varepsilon_r = 3.2$. The dimensional details of the antenna are depicted in Figure 6.23.

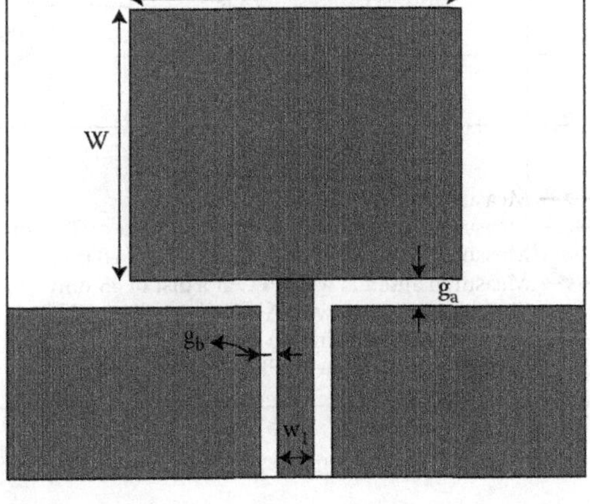

FIGURE 6.23
Dimensional details of the patch monopole antenna used in the study. Dimensions (mm): L = 18.8, W = 15.4, $g_a = 0.5$, $g_b = 0.14$, and $w_l = 1.83$.

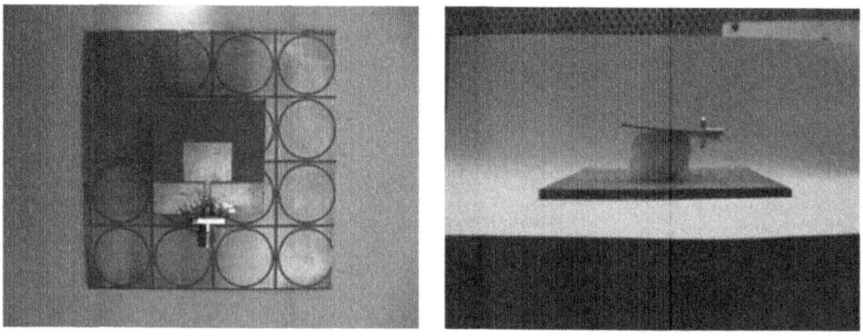

FIGURE 6.24
Photograph of the fabricated AMC-C with antenna on top: top view (a) and side view (b).

The AMC-C structure used was a 4×4 array (90×90 mm^2), and the PEC was 90×90 mm^2. The overall size of the antenna substrate was 34×34 mm^2. A photograph of the radiating system is shown in Figure 6.24. The antenna ground plane and the AMC-C ground plane are not connected. As can be seen, spacing between the antenna and the AMC is maintained by foam. The antenna, foam, and AMC structure are held together by cello tape. A coaxial cable connecting the antenna to a spectrum analyzer was fixed to the antenna rotator so as not to disturb the radiating system. The measurements were repeated several times to ensure the stability of the results.

The simulated and measured return loss S_{11} of the antenna in free space is compared with the return loss when the antenna is placed over the proposed structure and over a PEC in Figure 6.25 (the size of the PEC was same as that of the AMC-C structure). The distance between the antenna and AMC/PEC is optimized so as to maintain the frequency bandwidth of the original antenna.

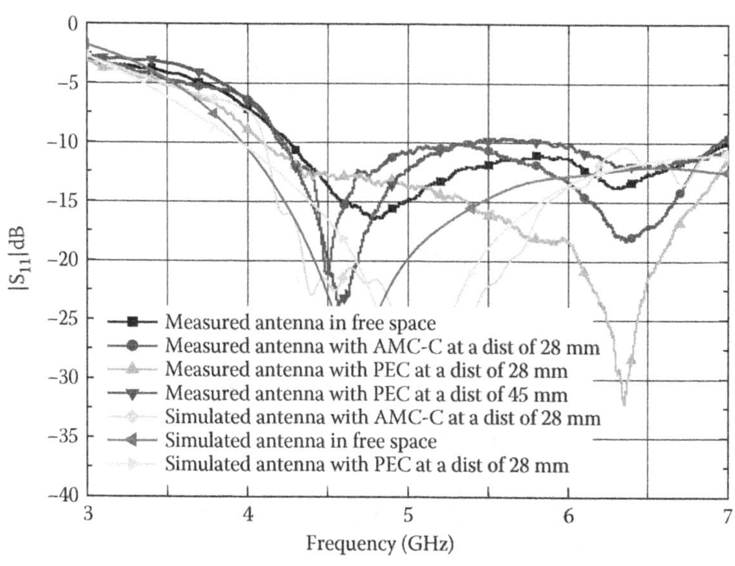

FIGURE 6.25
Measured return loss of the antenna in free space in comparison with the antenna–AMC and antenna–PEC systems.

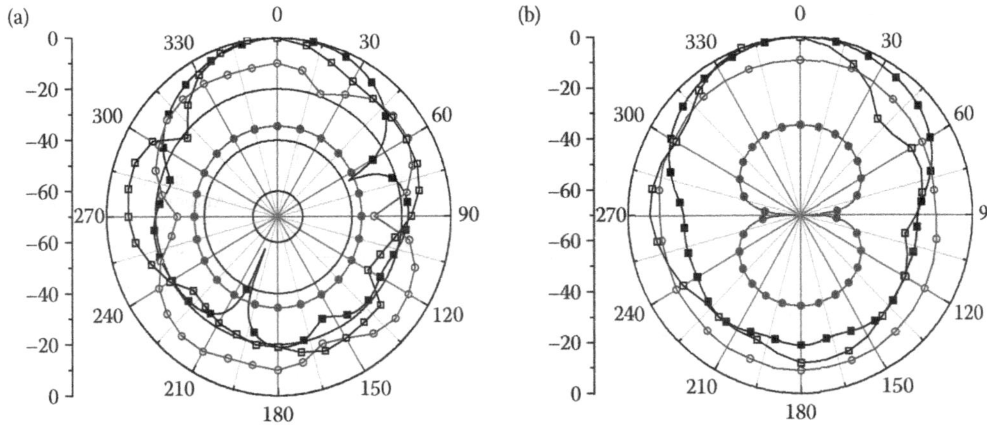

FIGURE 6.26
Measured radiation patterns in (a) E-plane and (b) H-plane. Rectangles represent the antenna with the AMC, and circles represent the antenna in free space. Solid symbols represent simulated results, and hollow symbols represent measured results.

It was observed experimentally that by adjusting the spacing between the antenna and the AMC, the impedance bandwidth of the original antenna can be altered, resulting in miniaturization. However, we chose the spacing so as to approximately maintain the original band of the antenna, as seen in Figure 6.24. It is observed that the required spacing is 28 mm with the AMC and 45 mm with the PEC. With this spacing, the overall size of the antenna–AMC system is $0.4\lambda_0$. The overall volume of the antenna radiating system is ≈225 cm^3.

The antenna is further characterized by its radiation properties and gain in an anechoic chamber. The E- and H-plane patterns at a randomly chosen frequency common to the AMC band and antenna (4.45 GHz) are plotted in Figure 6.26. As expected, with the use of the AMC, the FBR is improved by about 15 dB.

The gain measurements of the antenna in free space, the antenna–AMC system, and the antenna–PEC system in the E-plane are plotted in Figure 6.27 (similar results are observed for the H-plane). It is observed that the gain at the randomly chosen AMC band frequency (4.45 GHz) is significantly higher for the antenna–AMC system than for the monopole in free space or the antenna–PEC system. However, the gain outside of the AMC band frequency does not show such a significant increase. The maximum gain improvement is observed at the center of the AMC band (4.36 GHz), where the reflections are exactly in phase with the incident fields. This, in addition to the array-like effect caused by the cells of the AMC structure (which can be considered subradiators) help improve the overall gain of the radiating system considerably.

6.3.4 A Tunable HIS and Its Application to a Dual-Band Microstrip Antenna

Reconfigurable and tunable systems—systems that can be reconfigured in terms of operating frequency, functionality, and so on—are a trend in the communications industry today. Many of these systems require an antenna that can work at more than one frequency. There have been numerous reports in the open literature on reconfigurable antenna structures. Most of these employ devices such as PIN/Schottky diodes, field effect transistors (FETs), and radio frequency micro electro mechanical systems (RF MEMS) switches. Another approach for achieving tunability is by way of tunable HISs/AMCs [144]. Sievenpiper

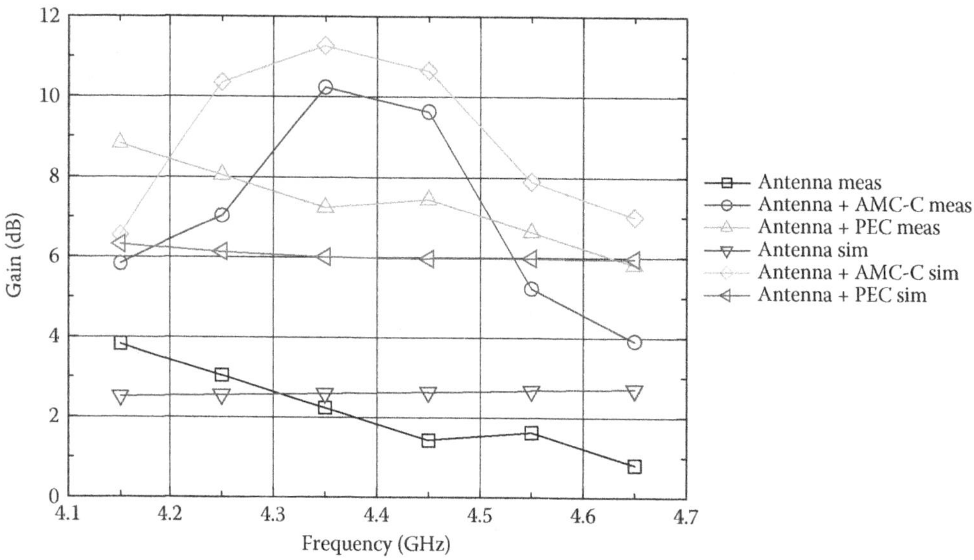

FIGURE 6.27
Comparison of measured gains for various structures.

reported a leaky wave antenna using a tunable HIS [145]. Other tunable antennas, including those that are HIS based, are reported in References 78, 86, 87, and 144–158. Most of these structures require a large number of active elements, which translates into added complexity of the biasing network and a high cost for the tunable antenna. Recently, attempts have been made to reduce the number of varactors in a tunable HIS. With the technique reported in Reference 159, the HIS and the varactors are on two different substrates, which may limit the practical usability of the tunable HIS. In this chapter, we report a simple method to tune a finite HIS using a varactor. The synthesized tunable HIS, when used as a ground plane for a microstrip antenna, results in resonance splitting. Each of these split bands can be tuned by varying the voltage applied to the varactor.

Several researchers have reported the use of EBG structure as an FSS. These structures have reflective characteristics and behave as AMCs over a band of frequencies (that is, the incident wave and reflected wave are in phase with each other over the HIS band). In this work, a standard square patch was taken as the starting point for the ease of fabrication as well as analysis. The unit cell dimensions of the HIS structure were optimized for an arbitrarily chosen frequency 5 GHz, and a probe-fed reference patch antenna was designed for resonance within the HIS band. A substrate with a dielectric constant 2.2 and thickness of 0.762 mm was used for both the patch antenna as well as the HIS layer. A 4×4 array of the HIS was placed below the microstrip antenna, thereby replacing the PEC ground by HIS/AMC ground. The microstrip antenna with HIS ground is shown in Figure 6.28a. The HIS band of the FSS was changed by incorporating shorting pins (or capacitors) between the HIS patches (Figure 6.28b).

To achieve tunability in the structure, the shorting pins were replaced with capacitors of different values, and it was observed that varying the capacitance from 0 to 2 pF varied the resonant frequency from 5 to 2 GHz, as shown in Figure 6.29. The shift in the AMC point caused a similar shift in the resonant frequency of the microstrip antenna. The simulated response of the antenna is shown in Figure 6.30. It is also seen from Figure 6.30 that the antenna is resonant at two frequencies: one below and one above the original frequency.

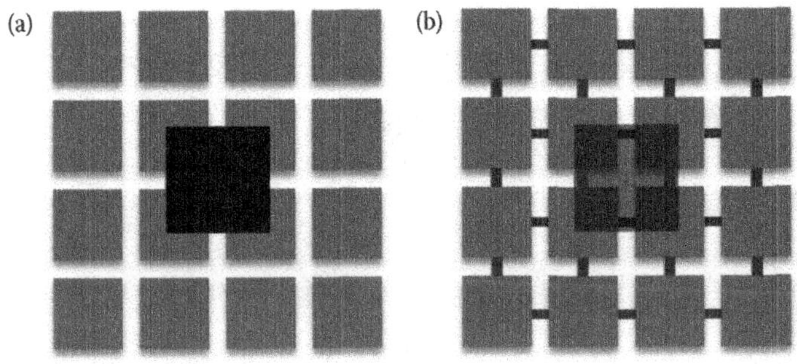

FIGURE 6.28
Patch antenna with an AMC surface as ground: (a) without shorting pins and (b) with shorting pins.

The probe-fed microstrip antenna design consists of a ground plane, an air gap of 60 mils to accommodate the varactor diodes, a single layer of Rogers RO4232 substrate, and the patch on top. The length and width of the patch are 15 and 21.5 mm, respectively, on a 30 mils thick substrate over a 120 × 120 mm (2λ × 2λ) ground plane. The reference antenna is designed to resonate at 5 GHz.

In order make the final design physically implementable at a very low cost, various simulations were carried out in CST Microwave Studio to study the effect of the number of elements in the AMC layer on the resultant frequencies of the antenna system. The current patterns on the AMC ground plane were studied in order to analyze the cause behind the shifting of resonant frequency. The complete analysis is discussed in subsequent sections. Figure 6.31 shows the current distribution on the 4 × 4 array of patches on the AMC ground. As is evident, only the four central HIS patches have the maximum current density. It was therefore hypothesized that eliminating the outer patches will not affect/degrade the performance of the antenna system. After further investigation, it was concluded that even a 2 × 2 array of patches in the HIS layer would be able to provide tunability, albeit with a slight reduction in the extent of the tuning.

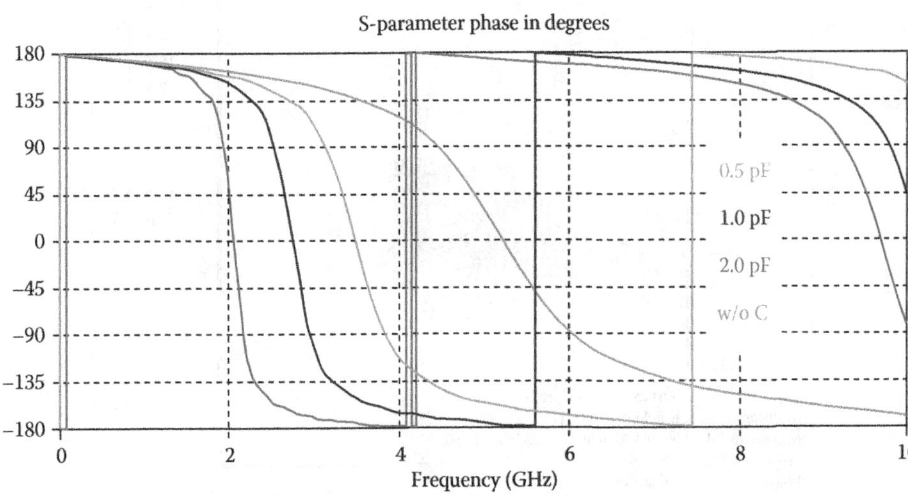

FIGURE 6.29
Reflection-phase characteristics of the AMC surface.

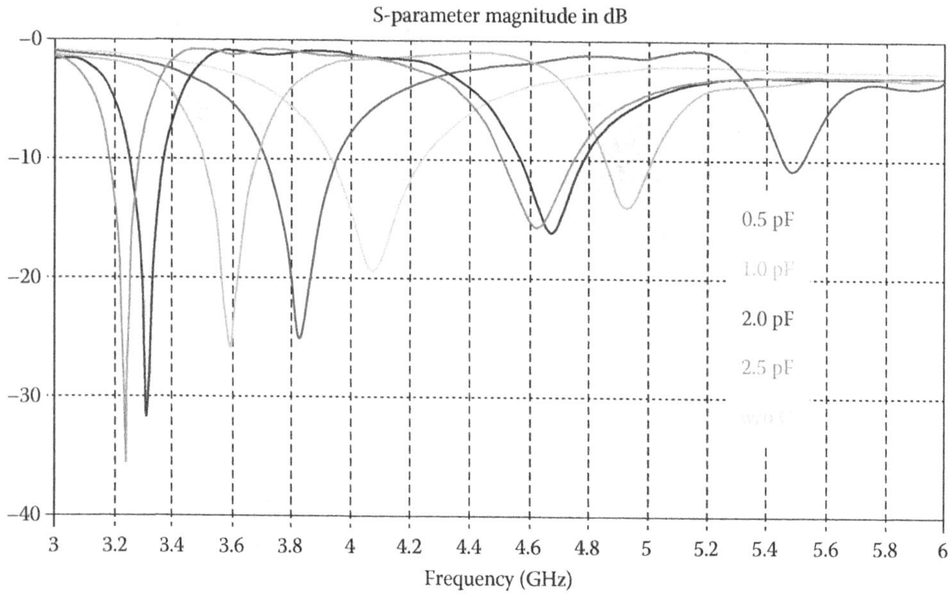

FIGURE 6.30
S_{11} for the patch antenna using capacitors instead of shorting pins.

Although the response of the antenna system was satisfactory in the desired band of tuning, there were additional resonances observed in the return loss at certain higher-frequency bands that were attributed to the patches in the AMC layer. In order to get these unwanted resonances out of the desired range, the thickness of the AMC was manipulated, and it was observed that with an increase in thickness, the unwanted resonances shifted

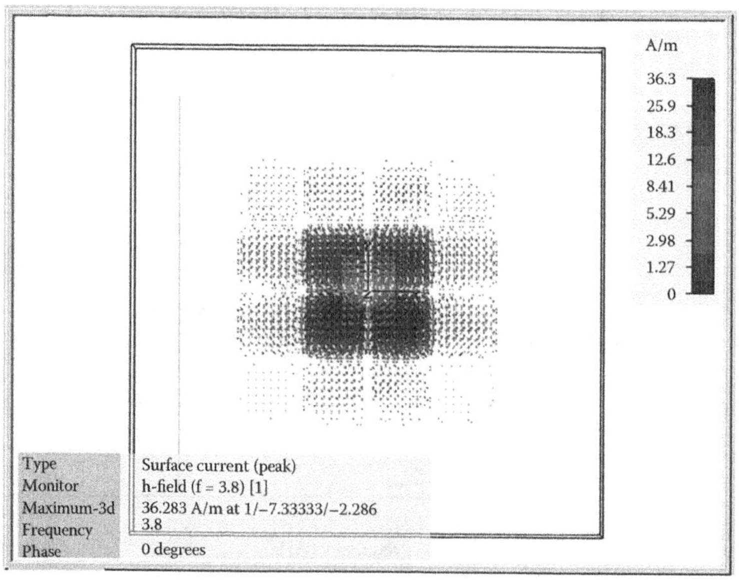

FIGURE 6.31
Current distribution on the AMC ground plane.

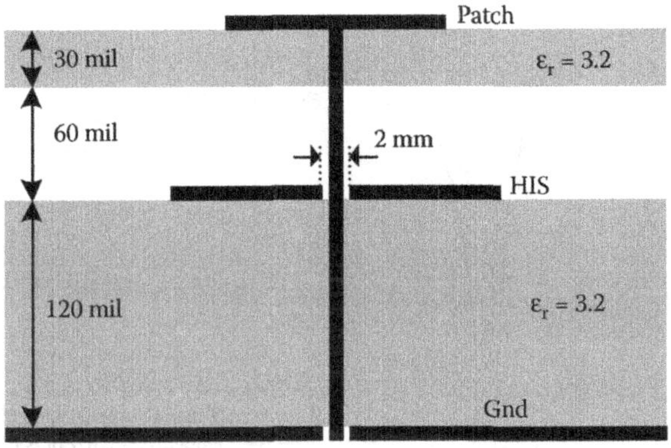

FIGURE 6.32
Cross section layout of the antenna system.

further outside of the desired band. The proposed AMC layer consists of a 2 × 2 matrix of metallic square patches on a 120 mil thick Rogers RO4232 layer, as shown in Figure 6.32. The HIS metal patches are of width 10 mm and periodicity 12 mm. Varactor diodes are mounted between the HIS patches in order to obtain the tuning in the patch antenna.

After verifying the above results using a fixed capacitor, the incorporation of varactor diodes was attempted. The biasing lines were modeled in the AMC plane to account for the direct current (DC) biasing. Hyperabrupt tuning varactor diodes SMV1232-079LF from Skyworks Solutions Inc. were used for the final design. The biasing circuit is comprised of two current-limiting 10 KΩ resistors, and the final layout is shown in Figure 6.33. The effect of the placement of the biasing circuit was investigated by carrying out both near-field and far-field measurements, and it was ascertained from these experiments that the results do not fluctuate much after the incorporation of the biasing circuit. The return loss S_{11} for the simulated and measured antenna with and without the biasing circuit is shown in Figure 6.34.

FIGURE 6.33
Fabricated AMC surface with biasing circuit.

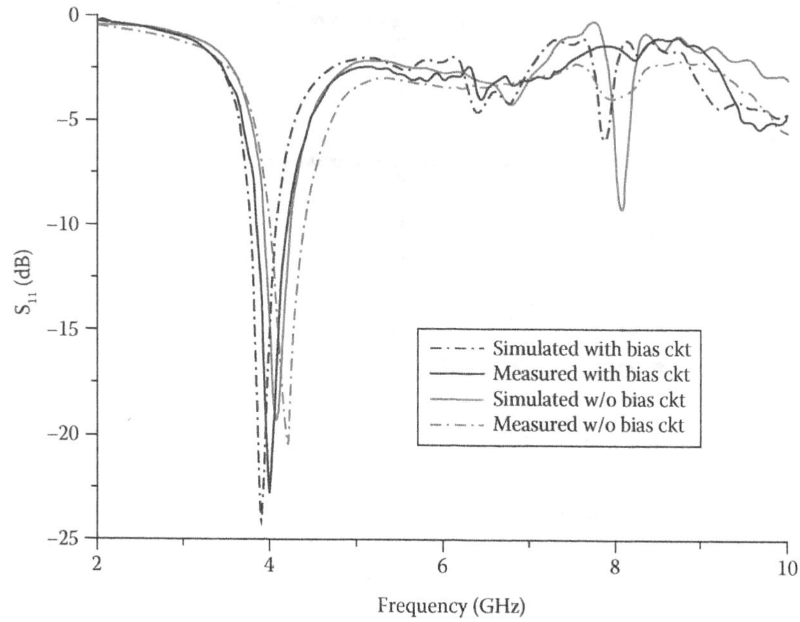

FIGURE 6.34
Return loss of the antenna with AMC surface and biasing circuit.

Further experimentation with the tunable antenna system revealed that only the capacitors across the radiating edges contribute to the shift in resonant frequency. This further simplified the circuit, and the AMC surface was slightly modified. The new AMC surface with biasing circuit is shown in Figure 6.35.

The measured results of the antenna system with diodes mounted on the AMC structure are shown in Figure 6.36a. The measured and simulated results are in agreement, as is evident from Figure 6.36b.

By comparing the radiation patterns at various frequencies, it was found that the patterns in the tuning range are remarkably stable and the 3 dB beam width remains constant. A comparison of the E- and H-plane patterns at various frequencies is presented in Figures 6.37 and 6.38, respectively.

FIGURE 6.35
Fabricated modified AMC surface with biasing circuit.

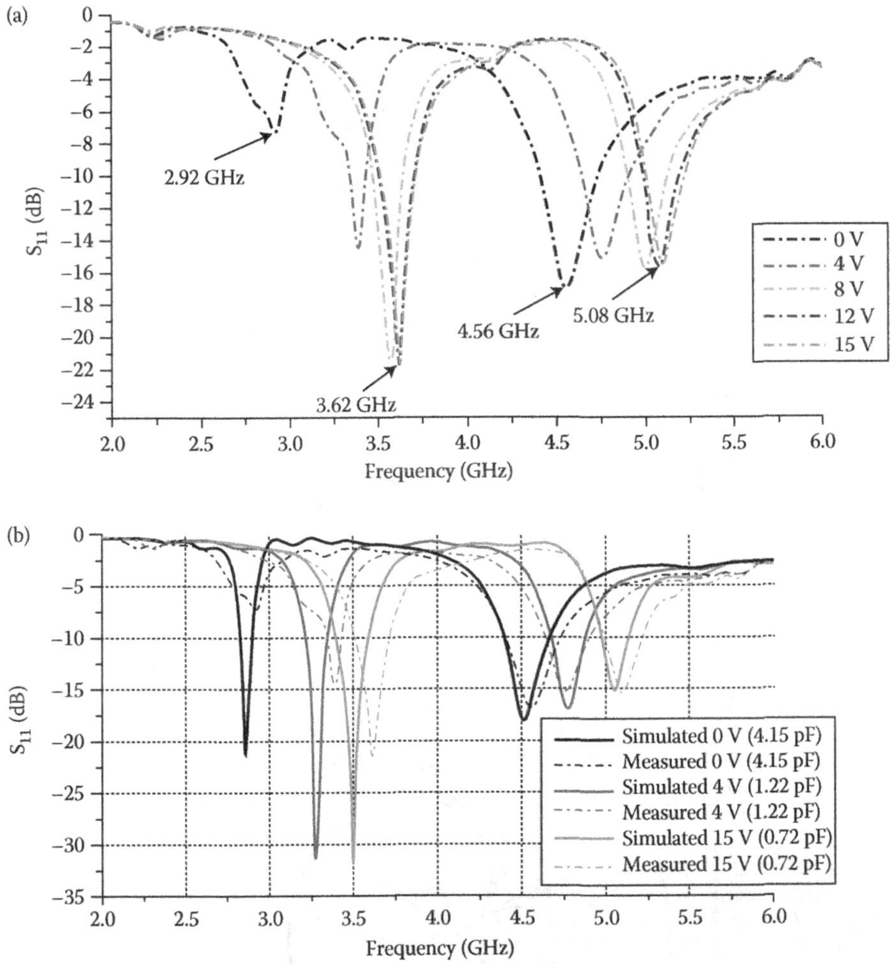

FIGURE 6.36
(a) Return loss of the reconfigurable antenna with varactor diodes. (b) Return loss of the fabricated antenna versus simulated results.

As is seen from Figure 6.34b, there is a considerable difference between the measured and simulated return loss at the lower frequency when the varactor capacitance is 4.15 pF. Moreover, the return loss at 2.9 GHz was found to be less than 10 dB, and the radiation pattern was also not in agreement with the simulated results. This difference was attributed to fabrication errors and the slight misplacement of the feed point. In order to keep the feed point at the same location and still change the feed point impedance, various antennas were fabricated with widths varying from 15 to 19 mm in steps of 2 mm. The antenna with a 15 mm width was found to have return loss better than 10 dB. The return loss at various varactor biasing voltages is shown in Figure 6.39. The radiation patterns at all frequencies were measured again and found to be similar to those obtained earlier, with an exception at capacitance 4.15 pF where, unlike in the previous case, the measured results matched the simulated ones. The variation in the feed point impedance resulted in the improvement of return loss beyond 10 dB. However, it did not provide the reason for such a discrepancy between the simulated and measured return loss at low frequency. The discrepancy was

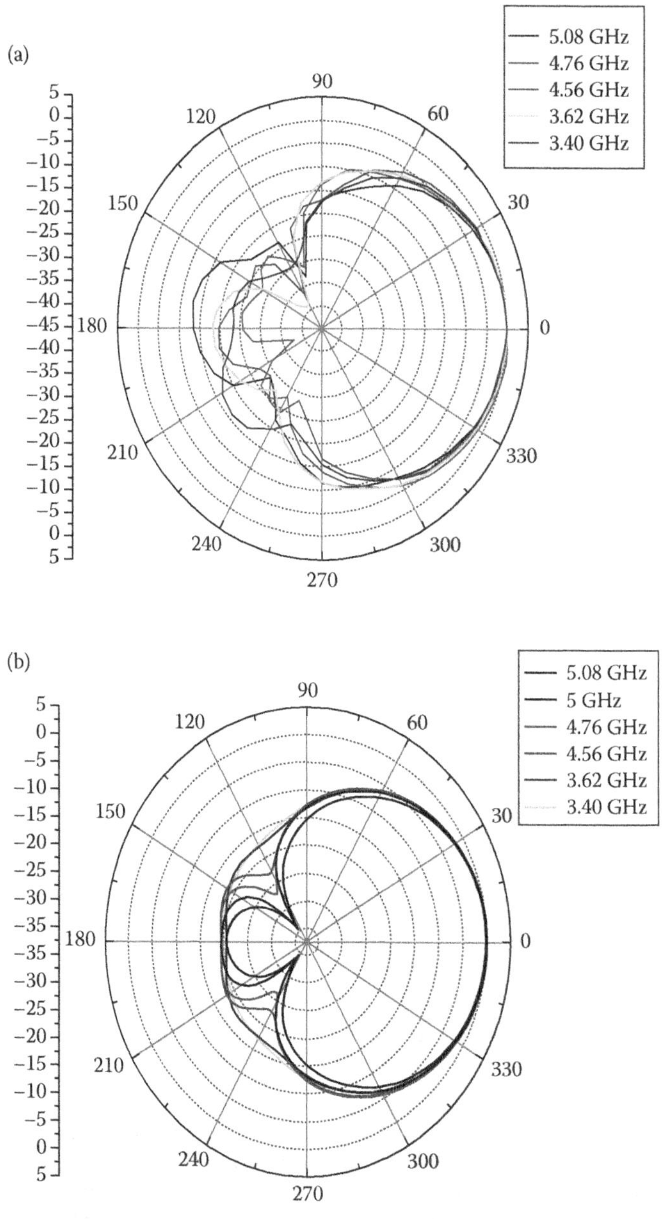

FIGURE 6.37
H-plane patterns of the antenna system: (a) measured and (b) simulated.

finally traced to the resistance offered by the varactor diode. The resistance of the varactor varies with the applied voltage as well as the frequency of operation.

The varactor resistance was therefore incorporated in the lumped component RLC model in CST Microwave Studio, and the results with additional resistance were compared with the measured results. The additional resistance of 1 Ohm was added to the simulation in order to account for resistance due to soldering and the increase in resistance at high frequency. The measured and simulated responses at 4.15 and 1.22 pF are shown in

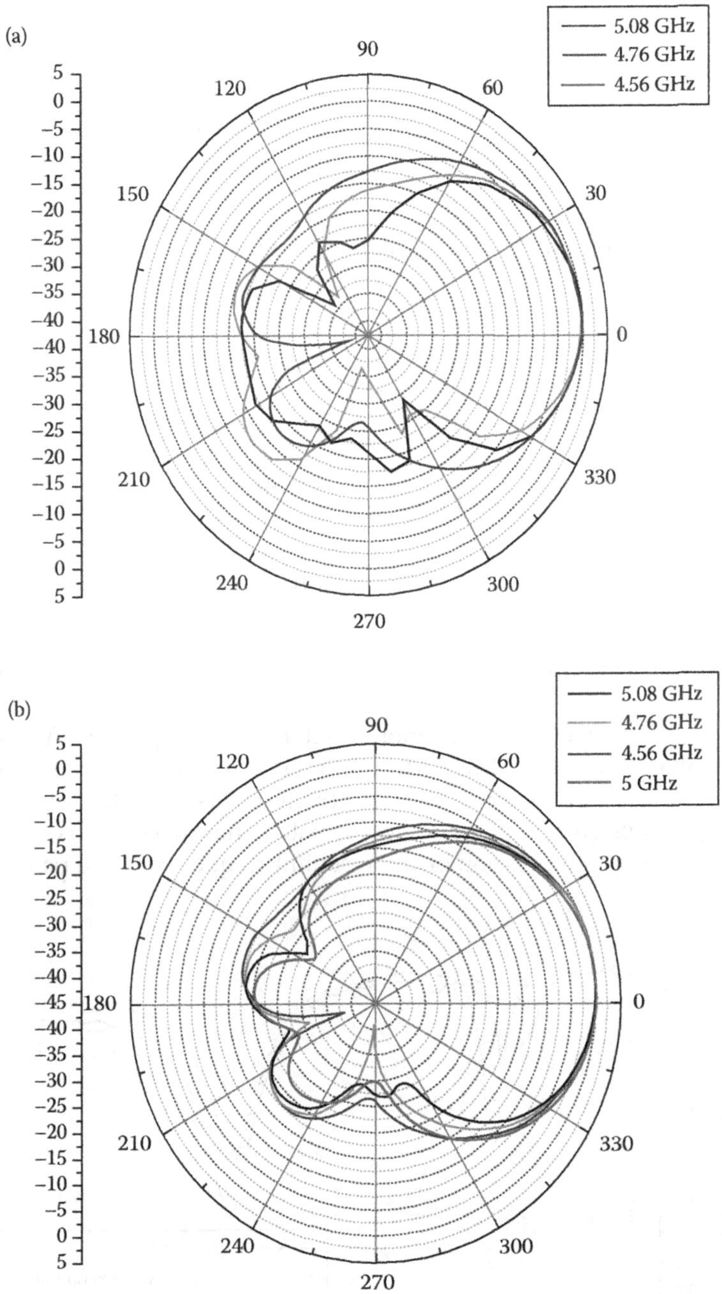

FIGURE 6.38
E-plane patterns of the antenna system: (a) measured and (b) simulated.

Figure 6.40. As can be seen from the return loss plot, there is now considerable agreement between the measured and simulated results.

The suppression of the original resonant frequency and the origin of two new frequencies were analyzed by monitoring the current and E-field patterns of the patch at new resonant frequencies f_{lower} and f_{upper}. The current and E-field patterns at the patch antenna's

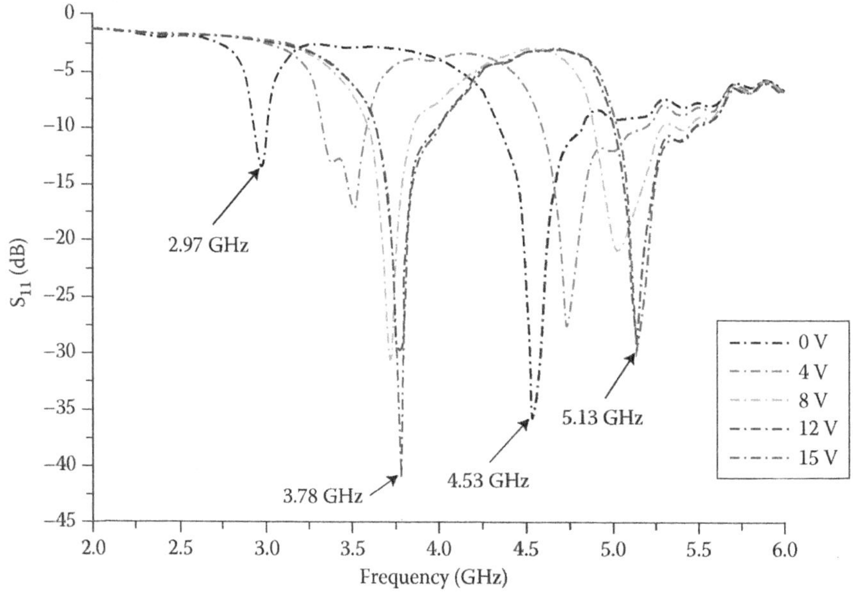

FIGURE 6.39
Return loss of the reconfigurable antenna with varactor diodes.

original resonant frequency were also monitored. For the purpose of analysis, the case of the reference antenna with the AMC surface and without a capacitor was used. In the present case, f_{lower} and f_{upper} are 4 and 8 GHz, respectively. It was found that, due to the presence of the high impedance ground plane below the patch, the currents at 5 GHz were severely diminished, leading to suppression of the resonant frequency of the patch antenna. The

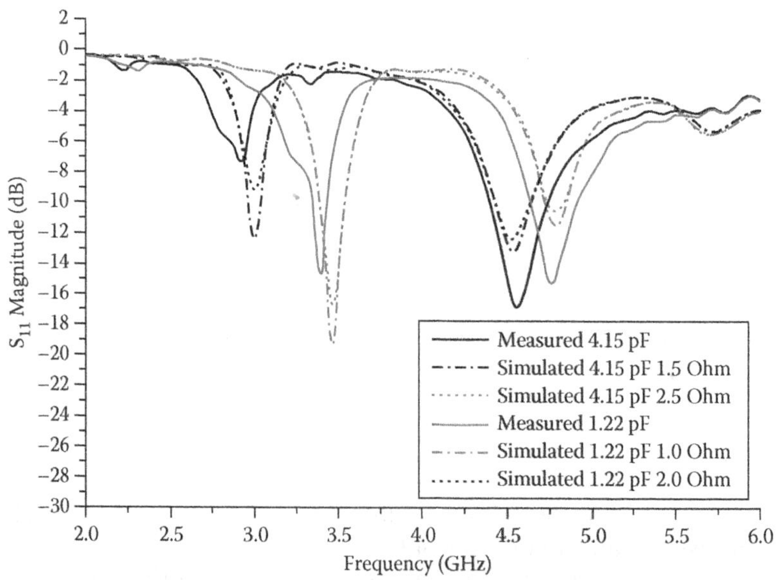

FIGURE 6.40
S_{11} of the antenna system after incorporating series resistance of the varactor.

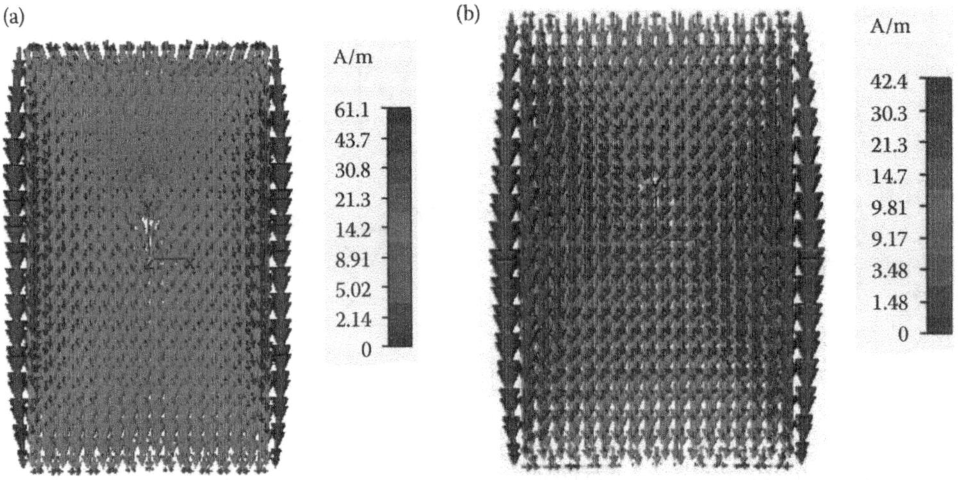

FIGURE 6.41
Current patterns on the patch. (a) 4 GHz. (b) 8 GHz.

current pattern over the patch at two frequencies was found to be identical. Using the cavity model and the analogy of magnetic current over an electric ground plane for a microstrip patch antenna, the E-field patterns were analyzed. It was found that the E-field patterns were almost identical, ruling out the possibility that two different modes were generated at the two frequencies. The current for the two frequencies is shown in Figure 6.41. The patterns show a variation of only one half cycle at both frequencies, thereby indicating TM_{10} mode at f_{lower} as well as f_{upper}.

In order to further analyze the origin of the two frequencies, the transmission line model and an equivalent RLC circuit for the antenna and the HIS were used. The transmission line equivalent circuit of the microstrip antenna is given in Reference 160, and that of an HIS has been given in References 161 and 162. These same models have been used for analysis of the structure and have been placed stacked one on top of other as shown in Figure 6.42. The HIS loaded with the varactor on top acts as a variable HIS, which can be used to tune the patch antenna.

The imaginary component of the earlier equivalent circuit's input admittance is given as

$$Y = \frac{-j[\omega^4(2l_2c_al_1c_1 + 2l_ac_al_2c_1 + l_ac_al_1c_1) - \omega^2(l_ac_a + 2l_2c_1 + 2l_2c_a + l_1c_1) + 1]}{\omega^5\left(l_al_1l_2c_1c_a + l_2^2l_1c_1c_a + l_2^2l_ac_1c_a\right) + \omega^3\left(l_al_1c_1 + l_2l_1c_1 + l_2^2c_1 + 2l_al_2c_a + l_2^2c_1\right) + \omega(l_a + l_2)} \tag{6.16}$$

Solving for ω at such a resonance that $Y = 0$, we get the expression for the frequency of the antenna system at different capacitance values in the AMC surface:

$$\omega = \sqrt{\frac{(l_ac_a + (2l_2 + l_1)c_1 + 2l_2c_a) \pm \sqrt{(l_ac_a + (2l_2 + l_1)c_1 + 2l_2c_a)^2 - 4(2l_2c_al_1c_1 + l_ac_a(2l_2 + l_1)c_1)}}{2(2l_2c_al_1c_1 + l_ac_a(2l_2 + l_1)c_1)}} \tag{6.17}$$

$$\omega = \sqrt{\frac{((1/\omega_a^2) + (1/\omega_h^2) + 2l_2c_a) \pm \sqrt{((1/\omega_a^2) + (1/\omega_h^2) + 2l_2c_a)^2 - 4(2l_2c_al_1c_1 + (1/\omega_a^2\omega_h^2))}}{2(2l_2c_al_1c_1 + (1/\omega_a^2\omega_h^2))}} \tag{6.18}$$

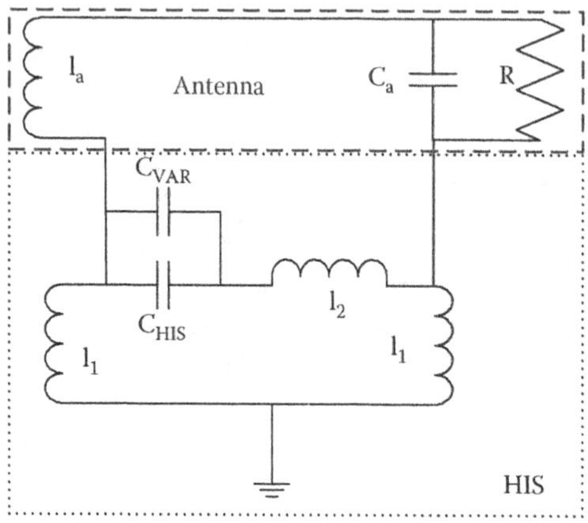

FIGURE 6.42
Equivalent circuit of the patch antenna and HIS.

where
ω = resonance frequency of the antenna system
ω_a = resonance frequency of the reference antenna
ω_h = center frequency of the AMC surface
l_a = inductance of the patch antenna
l_1 = inductance of the AMC surface
l_2 = inductance of the dielectric substrate
c_a = capacitance of the patch antenna
c_1 = capacitance of the AMC surface

Equation 6.18 proves the origin of two frequencies, f_{lower} and f_{upper}, and the suppression of the original resonant frequency of the microstrip patch antenna. Circuit parameters for the reference antenna were extracted using standard equations available in the literature [160]. The Z parameters extracted from CST Microwave Studio indicate the presence of significant probe inductance and capacitance. The inductance and capacitance were computed and found to be $L_{probe} = 1.9$ nH and $C_{probe} = 2.798$ pF.

The AMC structure simulated for reflection phase gives the response for an infinitely long HIS. Hence, the standard design equations cannot be used for computing the values of L and C of the AMC ground plane. Consequently, Equation 6.18 was used for computing the L and C values of the finite HIS embedded between two dielectrics. The equivalent admittance was set to zero in Equation 6.19 to obtain the frequency at which the structure will start resonating:

$$\omega^4(2l_2c_al_1c_1 + l_ac_a(2l_2 + l_1)c_1) - \omega^2(l_ac_a + (2l_2 + l_1)c_1 + 2l_2c_a) + 1 = 0 \quad (6.19)$$

Equation 6.19 has three unknowns: l_1, C_1 ($C_{HIS} + C_{var}$), and l_2. In light of this, one set of two frequencies generated by the placement of the varactor is not sufficient for solving the resulting simultaneous equations. Hence, two sets of frequencies obtained from an extreme set of capacitances were used to construct four equations from three unknowns. At $C_{var} = 4.15$ pF, the measured frequencies were 2.92 and 4.56 GHz, whereas at $C_{var} = 0.72$ pF, the measured frequencies were 3.62 and 5.08 GHz. Using this set of equations, the lumped element values for the finite HIS were determined to be $C_{HIS} = 3.0793$ pF, $l_1 = 0.2351$ nH,

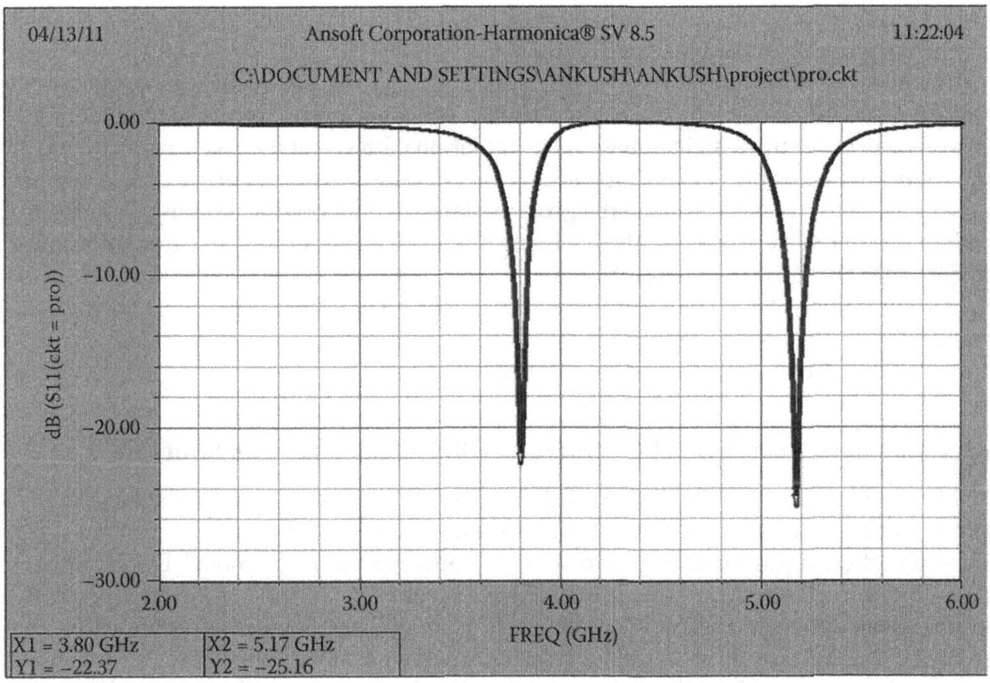

FIGURE 6.43
Simulated response in Serenade.

and $l_2 = 0.0684$ nH. The values were plugged into the equivalent circuit made in Serenade, and the resulting response was compared with the measured values. The return loss for varactor capacitance of 0.72 pF is shown in Figure 6.43.

Similar graphs were obtained for other values of the varactor capacitance. A comparison of predicted and actual frequencies was carried out, and the results are as shown in Table 6.2.

In conclusion, a reconfigurable antenna system using a finite AMC, which can be tuned to two different bands simultaneously, was proposed and verified experimentally in this chapter. Varactor diodes were successfully implemented, and the experimental results obtained were an excellent match for the simulated results. Utilizing a finite AMC ground

TABLE 6.2
Comparison of Calculated and Measured Results for Various Values of Varactor Capacitance

	Case I	Case II	Case III
C_{VAR} (pF)	0.72	1.22	4.15
f_{lower} predicted (GHz)	3.80	3.66	2.97
f_{lower} measured (GHz)	3.62	3.40	2.92
Offset (GHz)	+0.18	+0.26	0.05
f_{upper} predicted (GHz)	5.17	5.06	4.81
f_{upper} measured (GHz)	5.08	4.76	4.56
Offset (GHz)	+0.09	+0.30	+0.25

Note: $L_S = 0.235$ nH, $L_{HIS} = 0.068$ nH, and $C_{HIS} = 3.08$ pF.

plane facilitated a compact design and required only two varactor diodes, thereby reducing the complexity of the biasing circuit.

The lower band of tunability provided a reduction in the size of the patch, enabling the patch to resonate even at lengths as low as 0.35λ. This provided an effective reduction of 40% in the length of the microstrip antenna. The radiation patterns of the antenna were measured and found to be largely stable in the entire tuning range. This was attributed to the placement of the active components on the ground plane and not on the radiating element itself.

The antenna system was analyzed using the cavity model and the transmission line model, and an equivalent circuit of the structure was proposed. The lumped element equivalent values of the structure were extracted and used for verification of the results obtained. The equations proposed in this chapter satisfy the origin of two frequencies and the tunability due to variation in the varactor capacitance.

6.3.5 Compact Mushroom EBG Structure with Defected Ground Structure

In this section, a mushroom EBG-based structure with defected ground structure is presented, which can help reduce the harmonics in antenna applications. Figure 6.44 shows the schematic diagram for the unit cell of the conventional mushroom EBG. It consists of a 3 × 1.5 mm rectangular metal pad (black) on top of a 3.6 × 2.1 mm substrate (green) with relative permittivity $\varepsilon_r = 3.2$ and thickness 0.762 mm. The metal pad is connected to the ground plane on the bottom of the substrate through a via at the center of the substrate (seen in white in Figure 6.44a). When these EBG unit cells are arranged in a two-dimensional lattice, the field between the adjacent metal pads through the gap provides the capacitance. The metal pads are connected to each other on the ground through vias, and the current flowing between the pads results in inductance. The inductance and capacitance are in parallel, resulting in parallel resonance circuit, providing high impedance, and exhibiting a bandgap at the resonance frequency.

The proposed EBG is achieved by cutting slots in the ground plane of the conventional mushroom EBG structure. The top view of the proposed structure is the same as the view shown in Figure 6.44a. The bottom view of the proposed EBG structure and its cross-sectional view are shown in Figure 6.45.

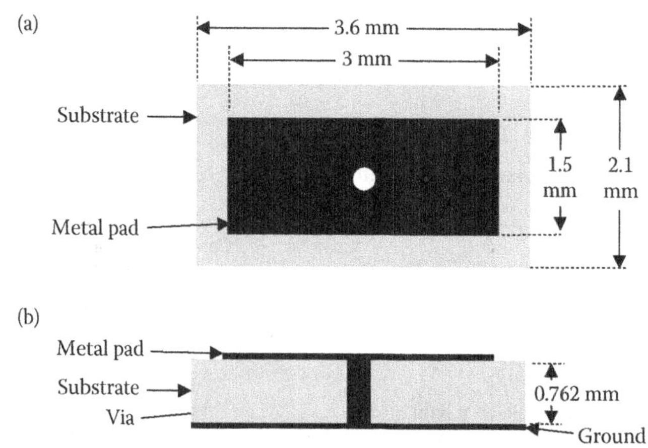

FIGURE 6.44
Conventional mushroom EBG: (a) top view and (b) cross-sectional view.

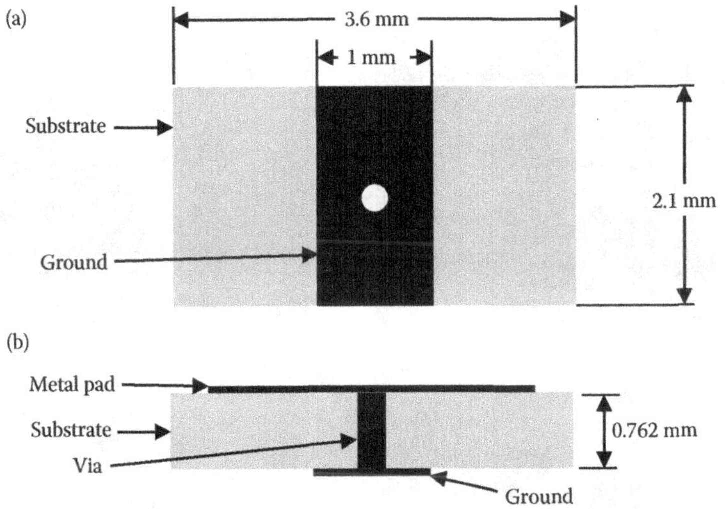

FIGURE 6.45
Proposed mushroom EBG: (a) bottom view and (b) cross-sectional view.

To characterize these EBG structures, the suspended microstrip line method is used. In this method, a transmission line of width 2.3 mm on a thin substrate is placed over a finite array of 7 × 7 mushroom EBG cells, as shown in Figure 6.46. The thin substrate is 0.127 mm thick with relative permittivity of 2.2. The transmission characteristics determine the characteristics of the EBG structure. Figure 6.46a shows the top view and Figure 6.46b shows the cross-sectional views of the measurement method with the conventional mushroom

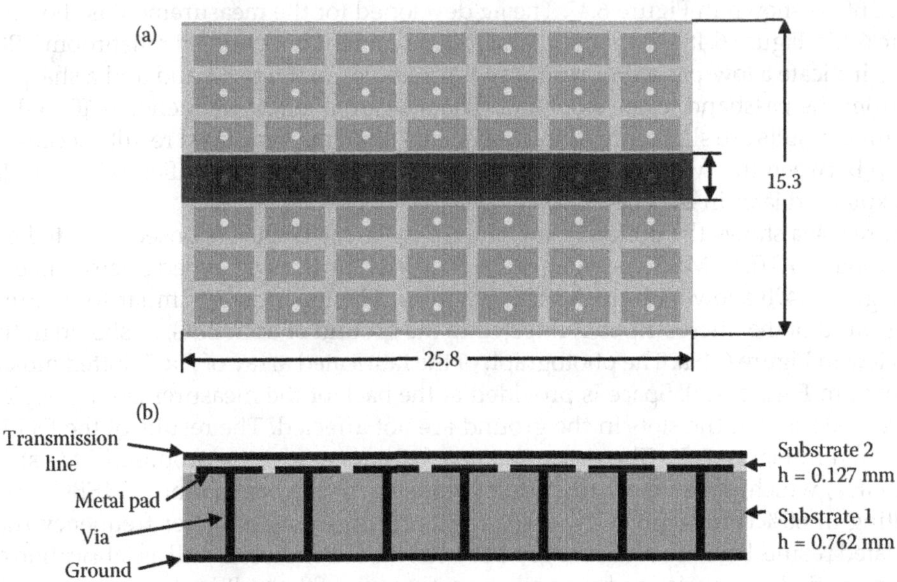

FIGURE 6.46
7 × 7 array of EBG cells characterized by the suspended microstrip method: (a) top view and (b) cross-sectional view.

(a) (b) (c)

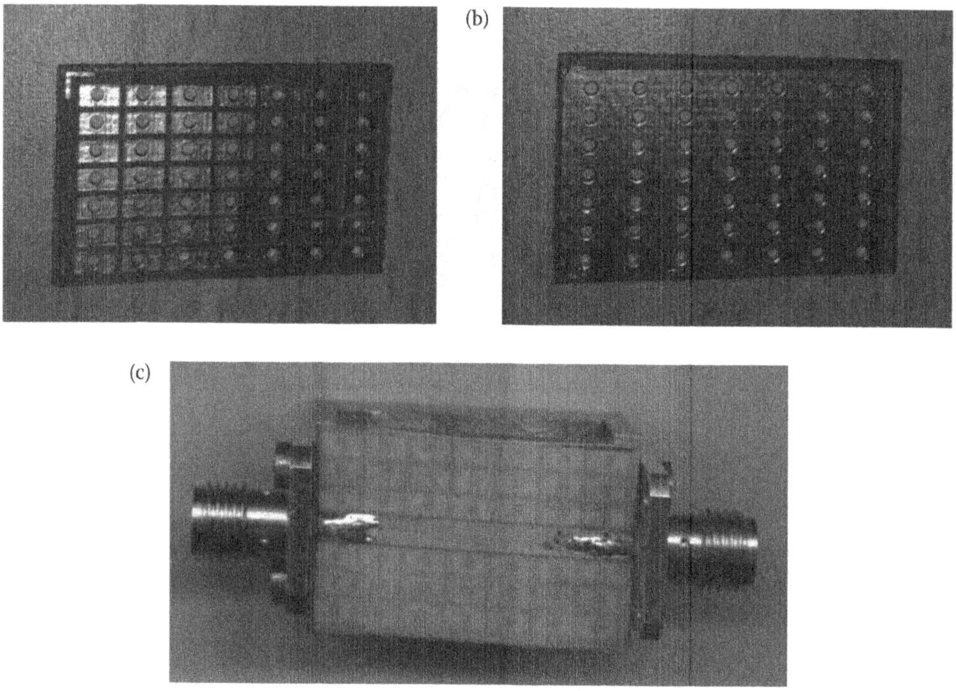

FIGURE 6.47
Photograph of the fabricated array of 7 × 7 conventional mushroom EBG structure: (a) top view, (b) bottom view, and (c) jig for measurement.

EBG (Figure 6.44). The photograph of the fabricated array of 7 × 7 conventional mushroom EBG is shown in Figure 6.47. The jig developed for the measurement is shown in the Figure 6.47c. Figure 6.48 shows the characteristics of the conventional mushroom EBG. The results indicate a low-pass filter response with ripples in the passband and a sharp transition from the passband to the stopband. The measured cutoff frequency is 10.4 GHz. The measured results are slightly higher in frequency than the simulated results because of the air gap between the two substrates in experimentation. This is verified with simulations and explained later in the chapter.

Figure 6.49a shows the bottom view of the 7 × 7 array of the proposed defected ground mushroom EBG (DG-MEBG) structure characterized by the suspended microstrip method, and Figure 6.49b shows the cross-sectional view. The top view is similar to Figure 6.46a. The ground of the microstrip is connected to the ground of each EBG, as shown in the bottom view in Figure 6.49a. The photograph of the fabricated array of 7 × 7 slotted mushroom is shown in Figure 6.50. Space is provided at the back of the measurement jig depicted in Figure 6.50c so that the slots in the ground are not affected. The results of the DG-MEBG are shown in Figure 6.51. The measured cutoff frequency of the proposed EBG structure is 4.2 GHz, which is far below the cutoff frequency of the conventional MEBG structure, resulting in 60% compaction. The measured results display a higher frequency than the simulated results because of the air gap between the substrates as well as fabrication errors. This is verified in simulation by creating an air gap of 80 μm. It is observed that with the air gap, the cutoff frequency shifts upward.

It can be seen from the Figure 6.49a that the slot in the ground makes the adjacent metal plates on top of the substrate connect through via with a thin line in the ground, which

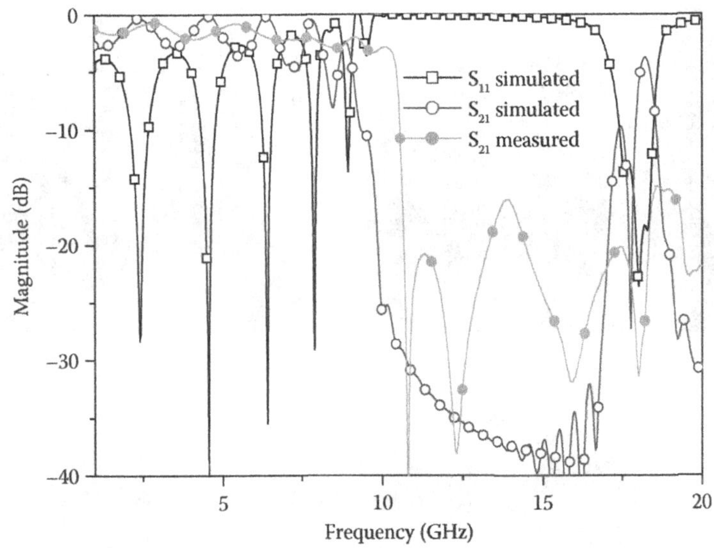

FIGURE 6.48
Measured and simulated results of the conventional mushroom EBG.

increases the inductance compared to a conventional MEBG. Since the slot in the ground plane of the proposed DG-MEBG is just below the gap between the adjacent metal plates on top of the substrate, as seen in Figure 6.49b, the field lines between the metal plate increase, resulting in an increase in the capacitance compared to a conventional MEBG. Increases in inductance and capacitance make the cutoff frequency shift downward and make the DG-MEBG more compact.

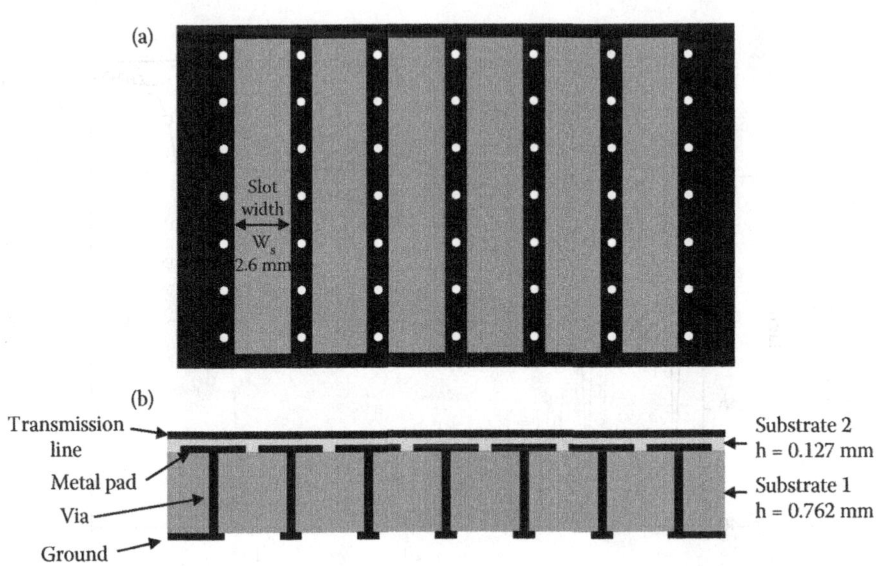

FIGURE 6.49
7 × 7 array of the proposed EBG structure characterized by the suspended microstrip method: (a) bottom view and (b) cross-sectional view.

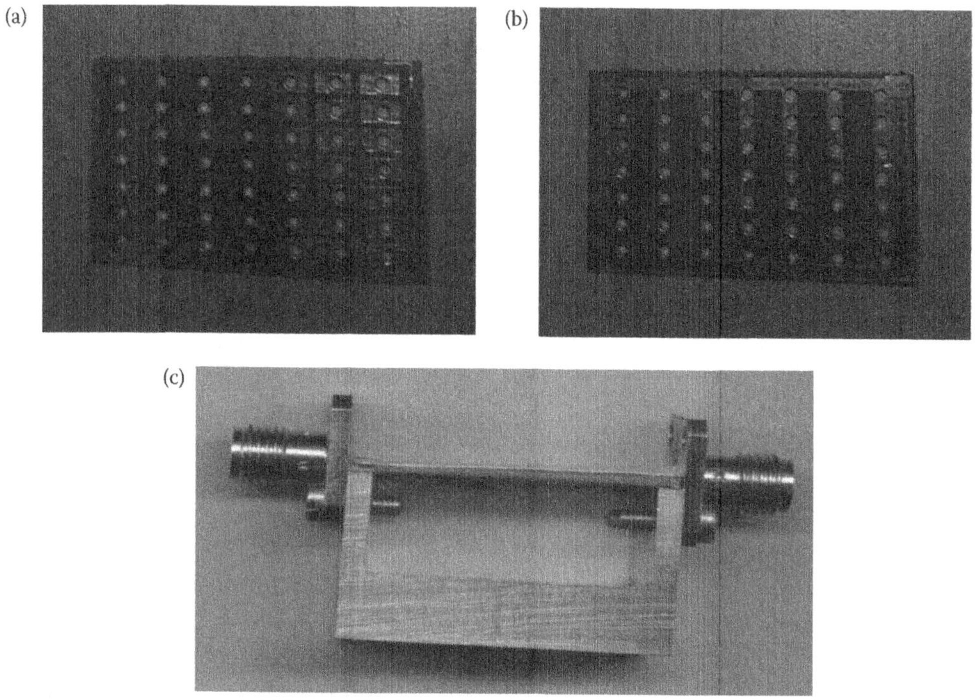

FIGURE 6.50
Photograph of the fabricated proposed array of 7 × 7 slotted mushroom EBG structure: (a) top view, (b) bottom view, and (c) jig for measurement.

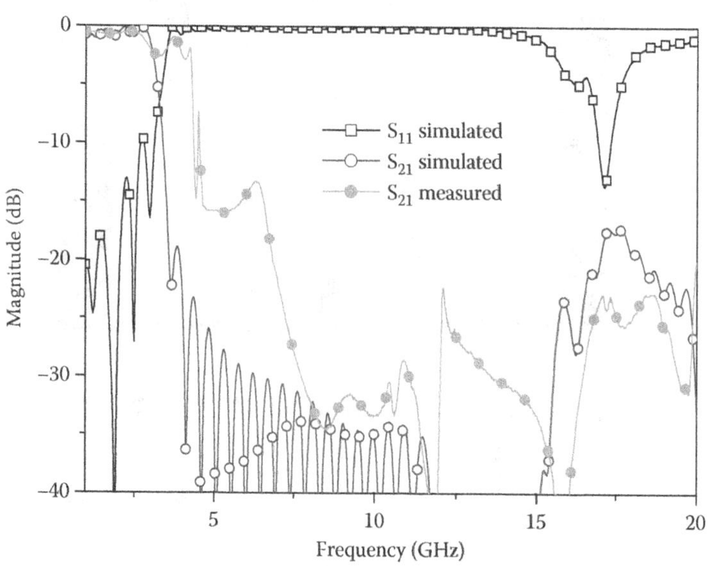

FIGURE 6.51
Measured and simulated results of the proposed mushroom EBG.

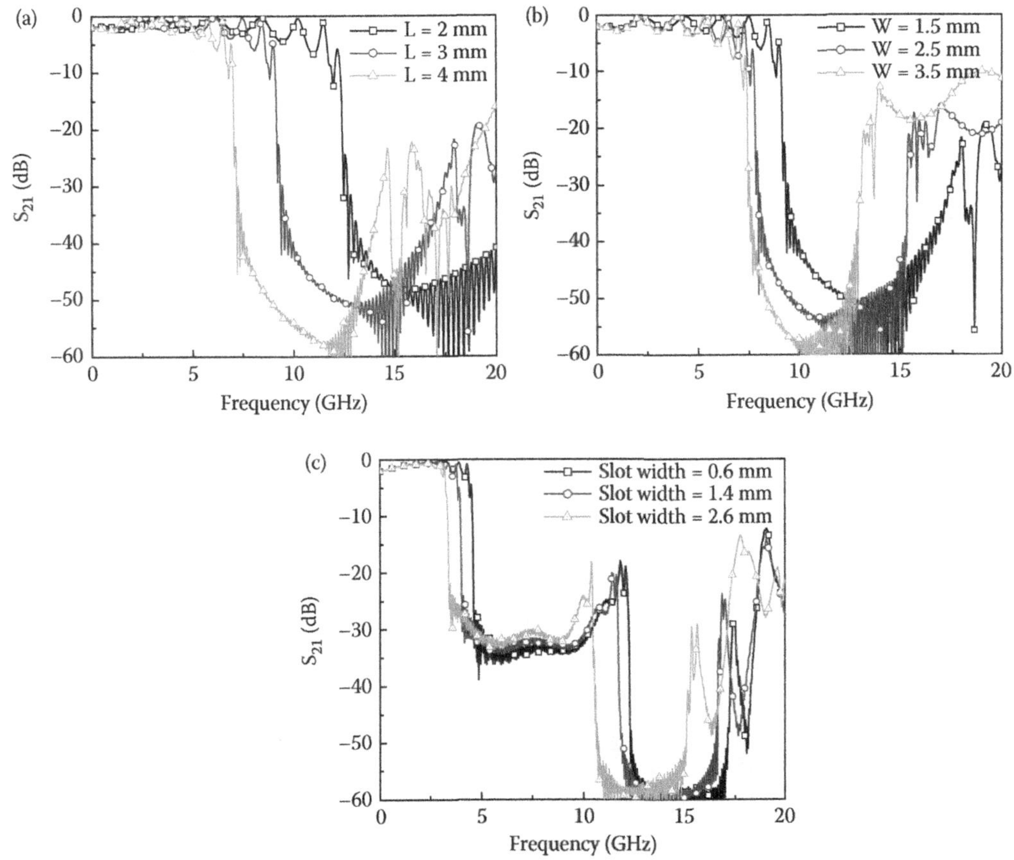

FIGURE 6.52
Simulated results with variations in the (a) rectangular patch length, (b) rectangular patch width, and (c) slot width of the proposed mushroom EBG.

A parametric analysis of the structure is carried out using CST Microwave Studio, and the simulated results are presented in Figure 6.52. It can be seen that with an increase in the length and width of the patch, the cutoff frequency shifts downward. The effect of a change in patch length is more than the effect of a change in width. An increase in the slot width in the ground plane causes the cutoff frequency to shift downward. The same phenomenon is observed when via diameter is decreased.

6.4 Conclusion

This chapter discussed the FSS properties of the proposed EBG cell. It was shown that the proposed FSS structure exhibits transmission characteristics at 10.8 GHz. A patch antenna was designed at 10.8 GHz and an FSS screen of a 13 × 13 array of unit cells was used as a superstrate to the antenna. A 7 dB improvement was seen in the broadside direction of the antenna. Other applications of periodic structures in antenna applications were also discussed.

References

1. F. Yang and Y. Rahmat-Samii, *Electromagnetic Band Gap Structures in Antenna Engineering*. Cambridge: Cambridge University Press, 2009.
2. E. Yablonovitch, Photonic band-gap structures, *J. Opt. Soc. Am. B*, vol. 10, no. 2, pp. 283–295, Feb. 1993.
3. J. D. Joannopoulos, S. G. Johnson, J. N. Winn, and R. D. Meade, *Photonic Crystals: Molding the Flow of Light*, 2nd Ed. Princeton, NJ: Princeton University Press, 2011.
4. J. Horak and Z. Raida, EBG substrates: Myths and reality, in *17th International Conference on Microwaves, Radar and Wireless Communications (MIKON)*, Wroclaw, Poland, May 19–21, 2008, pp. 1–4.
5. H.-Y. D. Yang, N. G. Alexopoulos, and E. Yablonovitch, Photonic band-gap materials for high-gain printed circuit antennas, *IEEE Trans. Antennas Propag.*, vol. 45, no. 1, pp. 185–187, Jan. 1997.
6. J. S. Colburn and Y. Rahmat-Samii, Patch antennas on externally perforated high dielectric constant substrates, *IEEE Trans. Antennas Propag.*, vol. 47, no. 12, pp. 1785–1794, Dec. 1999.
7. R. Gonzalo, P. de Maagt, and M. Sorolla, Enhanced patch-antenna performance by suppressing surface waves using photonic-bandgap substrates, *IEEE Trans. Microw. Theory Tech.*, vol. 47, no. 11, pp. 2131–2138, Nov. 1999.
8. Y. Qian, V. Radisic, and T. Itoh, Simulation and experiment of photonic band-gap structures for microstrip circuits, in *Asia-Pacific Microwave Conference Proceedings, (APMC)*, Hong Kong, Dec. 2–5, 1997, vol. 2, pp. 585–588.
9. D. Dey and R. S. Kshetrimayum, High gain and efficient patch antenna on micromachined GaAs EBGs with increased bandwidth, in *Annual IEEE India Conference*, Sep. 15–17, 2006, pp. 1–5.
10. A. Weily, L. Horvath, K. P. Esselle, B. C. Sanders, and T. S. Bird, A planar resonator antenna based on a woodpile EBG material, *IEEE Trans. Antennas Propag.*, vol. 53, no. 1, pp. 216–223, Jan. 2005.
11. D. Sievenpiper, L. Zhang, R. F. J. Broas, N. G. Alexópolous, and E. Yablonovitch, High-impedance electromagnetic surfaces with a forbidden frequency band, *IEEE Trans. Microw. Theory Tech.*, vol. 47, no. 11, pp. 2059–2074, Nov. 1999.
12. D. Sievenpiper, High-impedance electromagnetic surfaces, PhD dissertation. Dept. Elect. Eng., University of California at Los Angeles, Los Angeles, 1999.
13. F.-R. Yang, K.-P. Ma, Y. Qian, and T. Itoh, A uniplanar compact photonic-bandgap (UC-PBG) structure and its applications for microwave circuit, *IEEE Trans. Microw. Theory Tech.*, vol. 47, no. 8, pp. 1509–1514, Aug. 1999.
14. R. Coccioli, F.-R. Yang, K.-P. Ma, and T. Itoh, Aperture-coupled patch antenna on UC-PBG substrate, *IEEE Trans. Microw. Theory Tech.*, vol. 47, no. 11, pp. 2123–2130, Nov. 1999.
15. V. Radisic, Y. Qian, R. Coccioli, and T. Itoh, Novel 2-D photonic bandgap structure for microstrip lines, *IEEE Microw. Guid. Wave Lett.*, vol. 8, no. 2, pp. 69–71, Feb. 1998.
16. F. Falcone, T. Lopetegi, and M. Sorolla, 1-D and 2-D photonic bandgap microstrip structures, *Microw. Opt. Technol. Lett.*, vol. 22, no. 6, pp. 411–412, Sep. 1999.
17. R. S. Kshetrimayum, S. S. Karthikeyan, and D. Dey, Bandgap determination of triangular lattice EBGs in the ground plane, *AEU – Int. J. Electron. Commun.*, vol. 63, no. 8, pp. 699–702, Aug. 2009.
18. N. C. Karmakar and M. N. Mollah, Investigations into nonuniform photonic-bandgap microstripline low-pass filters, *IEEE Trans. Microw. Theory Tech.*, vol. 51, no. 2, pp. 564–572, Feb. 2003.
19. S. Y. Huang and Y. H. Lee, Tapered dual-plane compact electromagnetic bandgap microstrip filter structures, *IEEE Trans. Microw. Theory Tech.*, vol. 53, no. 9, pp. 2656–2664, Sep. 2005.
20. K.-P. Ma, K. Hirose, F.-R. Yang, Y. Qian, and T. Itoh, Realisation of magnetic conducting surface using novel photonic bandgap structure, *Electron. Lett.*, vol. 34, no. 21, pp. 2041–2042, Oct. 1998.
21. F.-R. Yang, Y. Qian, R. Coccioli, and T. Itoh, A novel low-loss slow-wave microstrip structure, *IEEE Microw. Guid. Wave Lett.*, vol. 8, no. 11, pp. 372–374, Nov. 1998.

22. M. K. Fries and R. Vahldierk, Small microstrip patch antenna using slow-wave structure, in *IEEE Antennas and Propagation Society International Symposium*, Salt Lake City, July 16–21, 2000, vol. 2, pp. 770–773.
23. L. Yang, M. Fan, and Z. Feng, A spiral electromagnetic bandgap (EBG) structure and its application in microstrip antenna arrays, in *Asia-Pacific Microwave Conference Proceedings (APMC)*, Suzhou, Dec. 4–7, 2005, vol. 3, 4 pp.
24. M. Hosseinipanah and Q. Wu, Miniaturised high-impedance surface with high angular stability of resonant frequency, *Electron. Lett.*, vol. 45, no. 24, pp. 1204–1206, Nov. 2009.
25. C. B. Mulenga and J. A. Flint, Planar electromagnetic bandgap structures based on polar curves and mapping functions, *IEEE Trans. Antennas Propag.*, vol. 58, no. 3, pp. 790–797, Mar. 2010.
26. Y. Fu, N. Yuan, and G. Zhang, Compact high-impedance surfaces incorporated with interdigital structure, *Electron. Lett.*, vol. 40, no. 5, pp. 310–311, Mar. 2004.
27. L. Yang, M. Fan, F. Chen, J. She, and Z. Feng, A novel compact electromagnetic-bandgap (EBG) structure and its applications for microwave circuits, *IEEE Trans. Microw. Theory Tech.*, vol. 53, no. 1, pp. 183–190, Jan. 2005.
28. E. Rajo-Iglesias, L. Inclan-Sanchez, J.-L. Vazquez-Roy, and E. Garcia-Munoz, Size reduction of mushroom-type EBG surfaces by using edge-located vias, *IEEE Microw. Wirel. Compon. Lett.*, vol. 17, no. 9, pp. 670–672, Sep. 2007.
29. L. Peng, C.-L. Ruan, and Z.-Q. Li, A novel compact and polarization-dependent mushroom-type EBG using CSRR for dual/triple-band applications, *IEEE Microw. Wirel. Compon. Lett.*, vol. 20, no. 9, pp. 489–491, Sep. 2010.
30. B.-Q. Lin, Q.-R. Zheng, and N.-C. Yuan, A novel planar PBG structure for size reduction, *IEEE Microw. Wirel. Compon. Lett.*, vol. 16, no. 5, pp. 269–271, May 2006.
31. Z. Z. Abidin, R. A. Abd-Alhameed, N. J. McEwan, S. M. R. Jones, K. N. Ramli, and A. G. Alhaddad, Design and analysis of UC-EBG on mutual coupling reduction, in *Loughborough Antennas Propagation Conference (LAPC)*, Loughborough, Nov. 16–17, 2009, pp. 693–696.
32. M. F. Abedin, M. Z. Azad, and M. Ali, Wideband smaller unit-cell planar EBG structures and their application, *IEEE Trans. Antennas Propag.*, vol. 56, no. 3, pp. 903–908, Mar. 2008.
33. K. Herbertz and S. Lucyszyn, Two-dimensional metamaterials for dual-band filter applications, in *38th European Microwave Conference (EuMC)*, Amsterdam, Oct. 28–30, 2008, pp. 1366–1369.
34. B.-Q. Lin, J. Liang, Y.-S. Zeng, and H.-M. Zhang, A novel compact and wide-band uni-planar EBG structure, *Prog. Electromagn. Res. C*, vol. 1, pp. 37–43, 2008.
35. W. Wang, X.-Y. Cao, W.-Y. Zhou, and T. Liu, A novel compact uni-planar electromagnetic bandgap (UC-EBG) structure, in *International Conference on Microwave and Millimeter Wave Technology (ICMMT)*, Nanjing, Apr. 21–24, 2008, vol. 4, pp. 1634–1636.
36. X. L. Bao, G. Ruvio, M. J. Ammann, and M. John, A novel GPS patch antenna on a fractal impedance surface substrate, *IEEE Antennas Wirel. Propag. Lett.*, vol. 5, no. 1, pp. 323–326, Dec. 2006.
37. F. Frezza, L. Pajewski, and G. Schettini, Fractal two-dimensional electromagnetic bandgap structures, *IEEE Trans. Microw. Theory Tech.*, vol. 52, no. 1, pp. 220–227, Jan. 2004.
38. M. N. Sujatha and K. J. Vinoy, Compact polarization dependent EBG surface with fractal boundary patches, in *IEEE Asia-Pacific Conference on Antennas and Propagation (APCAP)*, Singapore, Aug. 27–29, 2012, pp. 197–198.
39. S. Tse, Y. Hao, and C. Parini, Mushroom-like high-impedance surface (HIS) with slanted VIAS, in *Loughborough Antennas and Propagation Conference (LAPC)*, Loughborough, Apr. 2–3, 2007, pp. 309–312.
40. Y. Yao, X. Wang, and Z. Feng, A novel dual-band compact electromagnetic bandgap (EBG) structure and its application in multi-antennas, in *IEEE Antennas and Propagation Society International Symposium*, Albuquerque, July 9–14, 2006, pp. 1943–1946.
41. Y. Qian, D. Sievenpiper, V. Radisic, E. Yablonovitch, and T. Itoh, A novel approach for gain and bandwidth enhancement of patch antennas, in *IEEE Radio and Wireless Conference (RAWCON)*, Colorado Springs, Aug. 9–12, 1998, pp. 221–224.

42. F.-R. Yang, R. Coccioli, Y. Qian, and T. Itoh, PBG-assisted gain enhancement of patch antennas on high-dielectric constant substrate, in *IEEE Antennas and Propagation Society International Symposium*, Orlando, FL, July 11–16, 1999, vol. 3, pp. 1920–1923.
43. F. Yang and Y. Rahmat-Samii, Microstrip antennas integrated with electromagnetic band-gap (EBG) structures: A low mutual coupling design for array applications, *IEEE Trans. Antennas Propag.*, vol. 51, no. 10, pp. 2936–2946, Oct. 2003.
44. B. Mohajer-Iravani, S. Shahparnia, and O. M. Ramahi, Coupling reduction in enclosures and cavities using electromagnetic band gap structures, *IEEE Trans. Electromagn. Compat.*, vol. 48, no. 2, pp. 292–303, May 2006.
45. Z. Iluz, R. Shavit, and R. Bauer, Microstrip antenna phased array with electromagnetic bandgap substrate, *IEEE Trans. Antennas Propag.*, vol. 52, no. 6, pp. 1446–1453, June 2004.
46. H. S. Farahani, M. Veysi, M. Kamyab, and A. Tadjalli, Mutual coupling reduction in patch antenna arrays using a UC-EBG superstrate, *IEEE Antennas Wirel. Propag. Lett.*, vol. 9, pp. 57–59, 2010.
47. Y. Su, L. Xing, Z. Z. Cheng, J. Ding, and C. J. Guo, Mutual coupling reduction in microstrip antennas by using dual layer uniplanar compact EBG (UC-EBG) structure, in *International Conference on Microwave and Millimeter Wave Technology (ICMMT)*, Chengdu, May 8–11, 2010, pp. 180–183.
48. M. N. Sujatha and K. J. Vinoy, A stacked ring-patch artificial substrate for surface waves suppression and in phase reflection, in *5th International Conference on Industrial and Information Systems (ICIIS)*, Mangalore, July 29–Aug. 1, 2010, pp. 75–79.
49. D. Guha, S. Biswas, M. Biswas, J. Y. Siddiqui, and Y. M. M. Antar, Concentric ring-shaped defected ground structures for microstrip applications, *IEEE Antennas Wirel. Propag. Lett.*, vol. 5, no. 1, pp. 402–405, Dec. 2006.
50. F.-R. Yang, K.-P. Ma, Y. Qian, and T. Itoh, A novel TEM waveguide using uniplanar compact photonic-bandgap (UC-PBG) structure, *IEEE Trans. Microw. Theory Tech.*, vol. 47, no. 11, pp. 2092–2098, Nov. 1999.
51. F.-R. Yang, Y. Qian, and T. Itoh, A novel uniplanar compact PBG structure for filter and mixer applications, in *IEEE MTT-S International Microwave Symposium Digest*, Anaheim, June 13–19, 1999, vol. 3, pp. 919–922.
52. V. Radisic, Y. Qian, and T. Itoh, Broad-band power amplifier using dielectric photonic bandgap structure, *IEEE Microw. Guid. Wave Lett.*, vol. 8, no. 1, pp. 13–14, Jan. 1998.
53. S. Biswas, D. Guha, and C. Kumar, Control of higher harmonics and their radiations in microstrip antennas using compact defected ground structures, *IEEE Trans. Antennas Propag.*, vol. 61, no. 6, pp. 3349–3353, June 2013.
54. J. Wang, H. Zhang, W.-H. Chen, and C. Sheng, Design and application of a novel CB-CPW structure, *Prog. Electromagn. Res. M*, vol. 4, pp. 133–142, 2008.
55. F. Yang and Y. Rahmat-Samii, Reflection phase characterizations of the EBG ground plane for low profile wire antenna applications, *IEEE Trans. Antennas Propag.*, vol. 51, no. 10, pp. 2691–2703, Oct. 2003.
56. Z. Li, G. Wang, and Y. Cao, A low-profile equiangular spiral antenna using a novel EBG ground plane, in *7th International Symposium on Antennas, Propagation and EM Theory (ISAPE)*, Guilin, Oct. 26–29, 2006, pp. 1–3.
57. F. Yang, Y. Qian, and T. Itoh, Low-profile cavity-backed slot antenna using UC-PBG substrate, in *IEEE Antennas and Propagation Society International Symposium*, Salt Lake City, July 16-21, 2000, vol. 3, pp. 1796–1799.
58. C. Chang, Y. Qian, and T. Itoh, Analysis and applications of uniplanar compact photonic bandgap structures, *Prog. Electromagn. Res.*, vol. 41, pp. 211–235, 2003.
59. B. Mohajer-Iravani and O. M. Ramahi, Suppression of EMI and electromagnetic noise in packages using embedded capacitance and miniaturized electromagnetic bandgap structures with high-k dielectrics, *IEEE Trans. Adv. Packag.*, vol. 30, no. 4, pp. 776–788, Nov. 2007.
60. M.-S. Zhang, Y.-S. Li, C. Jia, and L.-P. Li, A power plane with wideband SSN suppression using a multi-via electromagnetic bandgap structure, *IEEE Microw. Wirel. Compon. Lett.*, vol. 17, no. 4, pp. 307–309, Apr. 2007.

61. M.-S. Zhang, Y.-S. Li, C. Jia, L.-P. Li, and J. Pan, A double-surface electromagnetic bandgap structure with one surface embedded in power plane for ultra-wideband SSN suppression, *IEEE Microw. Wirel. Compon. Lett.*, vol. 17, no. 10, pp. 706–708, Oct. 2007.
62. P. H. Rao and M. Swaminathan, A novel compact electromagnetic bandgap structure in power plane for wideband noise suppression and low radiation, *IEEE Trans. Electromagn. Compat.*, vol. 53, no. 4, pp. 996–1004, Nov. 2011.
63. P. H. Rao, Wideband electromagnetic band gap (EBG) structure for EMI suppression, in *10th International Conference on Electromagnetic Interference and Compatibility (INCEMIC)*, Bangalore, Nov. 26–27, 2008, pp. 13–15.
64. A. Foroozesh and L. Shafai, Investigation into the effects of the patch-type FSS superstrate on the high-gain cavity resonance antenna design, *IEEE Trans. Antennas Propag.*, vol. 58, no. 2, pp. 258–270, Feb. 2010.
65. A. Pirhadi, H. Bahrami, and J. Nasri, Wideband high directive aperture coupled microstrip antenna design by using a FSS superstrate layer, *IEEE Trans. Antennas Propag.*, vol. 60, no. 4, pp. 2101–2106, Apr. 2012.
66. E. Moharamzadeh and A. M. Javan, Triple-band frequency-selective surfaces to enhance gain of X-band triangle slot antenna, *IEEE Antennas Wirel. Propag. Lett.*, vol. 12, pp. 1145–1148, 2013.
67. Z.-H. Wu and W.-X. Zhang, Broadband printed compound air-fed array antennas, *IEEE Antennas Wirel. Propag. Lett.*, vol. 9, pp. 187–190, 2010.
68. Y.-J. Kim, S.-S. Nam, and H.-M. Lee, Frequency selective surface superstrate for wideband code division multiple access system, in *European Wireless Technology Conference (EuWIT 2009)*, Rome, Sep. 28–29, 2009, pp. 33–36.
69. L. Peng and C.-L. Ruan, UWB band-notched monopole antenna design using electromagnetic-bandgap structures, *IEEE Trans. Microw. Theory Tech.*, vol. 59, no. 4, pp. 1074–1081, Apr. 2011.
70. M. Yazdi and N. Komjani, Design of a band-notched UWB monopole antenna by means of an EBG structure, *IEEE Antennas Wirel. Propag. Lett.*, vol. 10, pp. 170–173, 2011.
71. M. Mokhtaari and J. Bornemann, Ultra-wideband microstrip antenna with coupled notch circuit, in *Proceedings of the 5th European Conference on Antennas and Propagation (EUCAP)*, Rome, Apr. 11–15, 2011, pp. 1435–1439.
72. T. Li, H.-Q. Zhai, G.-H. Li, and C.-H. Liang, Design of compact UWB band-notched antenna by means of electromagnetic-bandgap structures, *Electron. Lett.*, vol. 48, no. 11, pp. 608–609, May 2012.
73. B.-W. Liu, Y.-Z. Yin, Y. Yang, S.-H. Jing, and A.-F. Sun, Compact UWB bandpass filter with two notched bands based on electromagnetic bandgap structures, *Electron. Lett.*, vol. 47, no. 13, pp. 757–758, June 2011.
74. F. Wei, Q. Y. Wu, X.-W. Shi, and L. Chen, Compact UWB bandpass filter with dual notched bands based on SCRLH resonator, *IEEE Microw. Wirel. Compon. Lett.*, vol. 21, no. 1, pp. 28–30, Jan. 2011.
75. F. Wei, L. Chen, X.-W. Shi, and B. Liu, Compact UWB bandpass filter with tunable notch band based on folded SIR, *Electron. Lett.*, vol. 47, no. 22, pp. 1229–1230, Oct. 2011.
76. J. Ding, A harmonic suppression antenna using split ring resonators coupled with microstrip line, in *7th International Symposium on Antennas, Propagation and EM Theory (ISAPE)*, Guilin, Oct. 26–29, 2006, pp. 1–3.
77. T. Li, H. Zhai, G. Li, L. Li, and C. Liang, Compact UWB band-notched antenna design using interdigital capacitance loading loop resonator, *IEEE Antennas Wirel. Propag. Lett.*, vol. 11, pp. 724–727, June 2012.
78. J. Liang and H. Y. D. Yang, Microstrip patch antennas on tunable electromagnetic band-gap substrates, *IEEE Trans. Antennas Propag.*, vol. 57, no. 6, pp. 1612–1617, June 2009.
79. M. F. Karim, A. Liu, A. Alphones, and A. Yu, A novel reconfigurable filter using periodic structures, in *IEEE MTT-S International Microwave Symposium Digest*, June 11–16, 2006, pp. 943–946.
80. M. F. Karim, A. Q. Liu, A. Alphones, and A. B. Yu, MEMS-based tunable bandstop filter using electromagnetic bandgap (EBG) structures, in *Asia-Pacific Microwave Conference Proceedings (APMC)*, Suzhou, Dec. 4–7, 2005, vol. 3, 943–946.

81. M. G. Bray and D. H. Werner, A novel design approach for an independently tunable dual-band EBG AMC surface, in *IEEE Antennas and Propagation Society International Symposium*, Monterey, CA, June 20–25, 2004, vol. 1, pp. 289–292.
82. H.-J. Lee, K. L. Ford, and R. J. Langley, Independent multiband antenna tuning using an active EBG, in *Loughborough Antennas and Propagation Conference (LAPC)*, Loughborough, Mar. 17–18, 2008, pp. 141–144.
83. B. Elamaran, I.-M. Chio, L.-Y. Chen, and J.-C. Chiao, A beam-steerer using reconfigurable PBG ground plane, in *Microwave Symposium Digest. 2000 IEEE MTT-S International*, Boston, June 11–16, 2000, vol. 2, pp. 835–838.
84. M. G. Bray and D. H. Werner, A broadband open-sleeve dipole antenna mounted above a tunable EBG AMC ground plane, in *IEEE Antennas and Propagation Society International Symposium*, Monterey, CA, June 20–25, 2004, vol. 2, pp. 1147–1150.
85. H.-J. Lee, R. J. Langley, and K. L. Ford, Tunable active EBG, in *The Second European Conference on Antennas and Propagation (EuCAP 2007)*, Edinburgh, Nov. 11–16, 2007, pp. 1–4.
86. C. Mias and J. H. Yap, A varactor-tunable high impedance surface with a resistive-lumped-element biasing grid, *IEEE Trans. Antennas Propag.*, vol. 55, no. 7, pp. 1955–1962, July 2007.
87. D. F. Sievenpiper, J. H. Schaffner, H. J. Song, R. Y. Loo, and G. Tangonan, Two-dimensional beam steering using an electrically tunable impedance surface, *IEEE Trans. Antennas Propag.*, vol. 51, no. 10, pp. 2713–2722, Oct. 2003.
88. L. Inclán-Sánchez, E. Rajo-Iglesias, J.-L. Vázquez-Roy, O. Quevedo-Teruel, and S. E. Barbin, Reconfigurable unit cell for EBG and artificial surfaces based on hairpin resonators, in *IEEE MTT-S International Microwave Optoelectronics Conference (IMOC)*, Natal, Brazil, Oct. 29–Nov. 1, 2011, pp. 191–195.
89. O. Quevedo-Teruel, L. Inclán-Sánchez, J.-L. Vazquez-Roy, and E. Rajo-Iglesias, Compact reconfigurable planar EBGs based on short-circuited hairpin resonators, *IEEE Microw. Wirel. Compon. Lett.*, vol. 23, no. 9, pp. 462–464, Sep. 2013.
90. M. Al-Husseini, J. Costantine, C. G. Christodoulou, S. E. Barbin, A. El-Hajj, and K. Y. Kabalan, A reconfigurable frequency-notched UWB antenna with split-ring resonators, in *Asia-Pacific Microwave Conference Proceedings (APMC 2010)*, Yokohama, Dec. 7–10, 2010, pp. 618–621.
91. A. Ibáñez Loinaz and C. del Río Bocio, EBG at microwave frequency range: Bragg and/or resonant effect? *Microw. Opt. Technol. Lett.*, vol. 42, no. 5, pp. 383–385, Sep. 2004.
92. C. Caloz, C.-C. Chang, and T. Itoh, A novel anisotropic uniplanar compact photonic band-gap (UC-PBG) ground plane, in *31st European Microwave Conference (EuMC)*, London, Sep. 24–26, 2001, pp. 1–3.
93. C. Caloz and T. Itoh, Multilayer and anisotropic planar compact PBG structures for microstrip applications, *IEEE Trans. Microw. Theory Tech.*, vol. 50, no. 9, pp. 2206–2208, Sep. 2002.
94. N. Boisbouvier, A. Louzir, F. Bolzer, A.-C. Tarot, and K. Mahdjoubi, Compact EBG structure for slot-line printed devices, presented at the *27th ESA Antenna Technology Workshop on Innovative Periodic Antennas*, Santiago de Compostela, Mar. 9–11, 2004. Available at https://www.researchgate.net/publication/4088884_A_double_layer_EBG_Structure_for_slot-line_printed_devices
95. B. Mohajer-Iravani and O. M. Ramahi, Wideband circuit model for planar EBG structures, *IEEE Trans. Adv. Packag.*, vol. 33, no. 1, pp. 169–179, Feb. 2010.
96. B. Mohajer-Iravani, Electromagnetic interference reduction using electromagnetic bandgap structures in packages, enclosures, cavities, and antennas, PhD dissertation. Dept. Elect. Comput. Eng., The University of Maryland, College Park, 2007.
97. I. Bahl, Transmission lines and lumped elements, in *Microwave Solid State Circuit Design*, 2nd Ed. ed. I. Bahl and P. Bhatia, Hoboken, NJ: John Wiley & Sons Inc., 2003, pp. 25–77.
98. R. E. Chaddock, The application of lumped element techniques to high frequency hybrid integrated circuits, *Radio Electron. Eng.*, vol. 44, no. 8, pp. 414–420, Aug. 1974.
99. H. Greenhouse, Design of planar rectangular microelectronic inductors, *IEEE Trans. Parts Hybrids Packag.*, vol. 10, no. 2, pp. 101–109, June 1974.
100. D. M. Krafcsik and D. E. Dawson, A closed-form expression for representing the distributed nature of the spiral inductor, in *Microwave and Millimeter-Wave Monolithic Circuits*, 1986, vol. 86, pp. 87–92.

101. K. C. Gupta, R. Garg, I. Bahl, and P. Bhartia, *Microstrip Lines and Slotlines*, 2nd Ed. Norwood, MA: Artech House, Inc., 1996.
102. S. S. Bedair, On the odd-mode capacitance of the coupled microstrip lines, *IEEE Trans. Microw. Theory Tech.*, vol. 28, no. 11, pp. 1225–1227, Nov. 1980.
103. I. J. Bahl, *Lumped Elements for RF and Microwave Circuits*. Norwood, MA: Artech House, Inc., 2003.
104. S. S. Bedair, Characteristics of some asymmetrical coupled transmission lines (short paper), *IEEE Trans. Microw. Theory Tech.*, vol. 32, no. 1, pp. 108–110, Jan. 1984.
105. R. Ghatak, P. Sarkar, R. K. Mishra, and D. R. Poddar, A compact UWB bandpass filter with embedded SIR as band notch structure, *IEEE Microw. Wirel. Compon. Lett.*, vol. 21, no. 5, pp. 261–263, May 2011.
106. S. Pirani, J. Nourinia, and C. Ghobadi, Band-notched UWB BPF design using parasitic coupled line, *IEEE Microw. Wirel. Compon. Lett.*, vol. 20, no. 8, pp. 444–446, Aug. 2010.
107. Q. Li, Z.-J. Li, C.-H. Liang, and B. Wu, UWB bandpass filter with notched band using DSRR, *Electron. Lett.*, vol. 46, no. 10, pp. 692–693, May 2010.
108. A. Ali and Z. Hu, Metamaterial resonator based wave propagation notch for ultrawideband filter applications, *IEEE Antennas Wirel. Propag. Lett.*, vol. 7, pp. 210–212, 2008.
109. L. Zhu, S. Sun, and W. Menzel, Ultra-wideband (UWB) bandpass filters using multiple-mode resonator, *IEEE Microw. Wirel. Compon. Lett.*, vol. 15, no. 11, pp. 796–798, Nov. 2005.
110. S. Sun and L. Zhu, Capacitive-ended interdigital coupled lines for UWB bandpass filters with improved out-of-band performances, *IEEE Microw. Wirel. Compon. Lett.*, vol. 16, no. 8, pp. 440–442, Aug. 2006.
111. R. D. Gupta, M. P. Abegaonkar, A. Basu, and S. K. Koul, Studies on ultrawideband filters using stub-loaded circular and triangular resonators, *Microw. Opt. Technol. Lett.*, vol. 50, no. 11, pp. 2917–2922, Nov. 2008.
112. Y. Sung, UWB monopole antenna with two notched bands based on the folded stepped impedance resonator, *IEEE Antennas Wirel. Propag. Lett.*, vol. 11, pp. 500–502, May, 2012.
113. Y.-H. Chun, H. Shaman, and J.-S. Hong, Switchable embedded notch structure for UWB bandpass filter, *IEEE Microw. Wirel. Compon. Lett.*, vol. 18, no. 9, pp. 590–592, Sep. 2008.
114. K. Rabbi and D. Budimir, Highly selective reconfigurable filter for UWB systems, *IEEE Microw. Wirel. Compon. Lett.*, vol. 24, no. 3, pp. 146–148, Mar. 2014.
115. K. Rabbi, L. Athukorala, C. Panagamuwa, J. C. Vardaxoglou, and D. Budimir, Compact UWB bandpass filter with reconfigurable notched band, *Electron. Lett.*, vol. 49, no. 11, pp. 709–711, May 2013.
116. K. Thirumalaivasan and R. Nakkeeran, Development of UWB bandpass filter with reconfigurable notches to suppress IEEE 802.11 a/b services, *Microw. Opt. Technol. Lett.*, vol. 54, no. 10, pp. 2423–2426, Oct. 2012.
117. S.-G. Mao and Y.-Z. Chueh, Coplanar waveguide bandpass filters with compact size and wide spurious-free stopband using electromagnetic bandgap resonators, *IEEE Microw. Wirel. Compon. Lett.*, vol. 17, no. 3, pp. 181–183, Mar. 2007.
118. H. Kuan, Y.-L. Lin, R.-Y. Yang, and Y.-C. Chang, A multilayered parallel coupled microstrip bandpass filter with embedded SIR cells to have a broad upper rejection band, *IEEE Microw. Wirel. Compon. Lett.*, vol. 20, no. 1, pp. 25–27, Jan. 2010.
119. C.-H. Ahn, D.-J. Jung, and K. Chang, Compact parallel-coupled line bandpass filter using double complementary split ring resonators, *Microw. Opt. Technol. Lett.*, vol. 55, no. 3, pp. 506–509, Mar. 2013.
120. B. Strassner and K. Chang, Wide-band low-loss high-isolation microstrip periodic-stub diplexer for multiple-frequency applications, *IEEE Trans. Microw. Theory Tech.*, vol. 49, no. 10, pp. 1818–1820, Oct. 2001.
121. T. Yang, P.-L. Chi, and T. Itoh, High isolation and compact diplexer using the hybrid resonators, *IEEE Microw. Wirel. Compon. Lett.*, vol. 20, no. 10, pp. 551–553, Oct. 2010.
122. H. Y. Zeng, G. M. Wang, D. Z. Wei, and Y. W. Wang, Planar diplexer using composite right-/left-handed transmission line under balanced condition, *Electron. Lett.*, vol. 48, no. 2, pp. 104–106, Jan. 2012.
123. D. M. Pozar, *Microwave Engineering*, 4th Ed. Hoboken, NJ: John Wiley and Sons, Inc., 2012.

124. R.-R. Xu, H.-C. Zhao, Z.-Y. Zong, and W. Wu, Dual-band capacitive loaded frequency selective surfaces with close band spacing, *IEEE Microw. Wirel. Compon. Lett.*, vol. 18, no. 12, pp. 782–784, Dec. 2008.
125. C. A. Balanis, *Antenna Theory: Analysis and Design*, 3rd Ed. New Delhi: Wiley India Pvt. Limited, 2005.
126. F. Yang and Y. Rahmat-Samii, Reflection phase characteristics of the EBG ground plane for low profile wire antennas, *IEEE Trans. Antennas Propag.*, vol. 51, no. 1, pp. 2691–2703, Oct. 2003.
127. H. Nakano, K. Hitosugi, N. Tatsuzawa, D. Togashi, H. Mimaki, and J. Yamauchi, Effects on the radiation characteristics of using a corrugated reflector with a helical antenna and an electromagnetic band-gap reflector with a spiral antenna, *IEEE Trans. Antennas Propag.*, vol. 53, no. 1, pp. 191–199, Jan. 2005.
128. H. Nakano, K. Kikkawa, N. Kondo, Y. Iitsuka, and J. Yamauchi, Low profile equiangular spiral antenna backed by an EBG reflector, *IEEE Trans. Antennas Propag.*, vol. 57, no. 5, pp. 1309–1318, May 2009.
129. H. Nakano, Y. Asano, G. Tsutsumi, and J. Yamauchi, A low-profile inverted F element array backed by an EBG reflector, in *Proc. IEEE Int. Symp. on Antennas and Propag.*, Albuquerque, July 9–14, 2006, pp. 2985–2988.
130. D. Qu, L. Shafai, and A. Foroozesh, Improving microstrip patch antenna performance using EBG substrates, *Inst. Elect. Eng. Proc.-Microw. Antennas. Propag.*, vol. 153, no. 6, pp. 558–563, Dec. 2006.
131. A. Foroozesh and L. Shafai, Application of combined electric- and magnetic-conductor ground planes for antenna performance enhancement, *Can J. Elect. Comput. Eng.*, vol. 33, no. 2, pp. 87–98, Spring, 2008.
132. A. Foroozesh and L. Shafai, Performance enhancement of a microstrip patch antenna using high impedance surfaces and different ground plane sizes, in *Proc. Int. Symp. of Antennas and Propagation, ISAP2007*, Aug. 20–24, Niigata, Japan, pp. 1158–1161.
133. L. Akhoondzadeh-Asl, D. J. Kern, P. S. Hall, and D. H. Werner, Wideband dipoles on electromagnetic bandgap ground planes, *IEEE Trans. Antennas Propag.*, vol. 55, no. 9, pp. 2426–2434, Sep. 2007.
134. S. R. Best and D. L. Hanna, Design of a broadband dipole in close proximity to an EBG ground plane, *IEEE Antennas Propag. Mag.*, vol. 50, no. 6, pp. 52–64, Dec. 2008.
135. N. A. Abbasi and R. J. Langley, Multiband-integrated antenna/artificial magnetic conductor, *IET Microw. Antennas Propag.*, vol. 5, no. 6, pp. 711–717, Apr. 2011.
136. H. R. Raad, A. I. Abbosh, H. M. Al-Rizzo, and D. G. Rucker, Flexible and compact AMC based antenna for telemedicine applications, *IEEE Trans. Antennas Propag.*, vol. 61, no. 2, pp. 524–531, Feb. 2013.
137. F. Costa, A. Monorchio, and G. Manara, Ultra-thin absorbers by using high impedance surfaces with resistive frequency selective surfaces, *2007 IEEE Antennas and Propagation Society International Symposium*, Honolulu, June 9–15, 2007, pp. 861–864.
138. A. E. Martynyuk, J. I. M. Lopez, and N. A. Martynyuk, Active frequency-selective surfaces based on loaded ring slot resonators, *Electron. Lett.*, vol. 41, no. 1, pp. 2–4, Jan. 2005.
139. D. Cure, T. M. Weller, and F. A. Miranda, Study of a low-profile 2.4-GHz planar dipole antenna using a high-impedance surface with 1-D varactor tuning, *IEEE Trans. Antennas Propag.*, vol. 61, no. 2, pp. 506–515, Feb. 2013
140. H.-J. Lee, K. L. Ford, and R. J. Langley, Independently tunable low-profile dual-band high-impedance surface antenna system for applications in UHF band, *IEEE Trans. Antennas Propag.*, vol. 60, no. 9, pp. 4092–4101, Sep. 2012.
141. M. Li and N. Behdad, Frequency selective surfaces for pulsed high-power microwave applications, *IEEE Trans. Antennas Propag.*, vol. 61, no. 2, pp. 677–687, Feb. 2013.
142. M. E. De Cos, Y. Álvarez, and F. Las-Heras, Planar artificial magnetic conductor: Design and characterization setup in the RFID SHF band, *J. Electromagn. Waves Appl.*, vol. 23, no. 11–12, pp. 1467–1478, 2009.
143. C. T. P. Song, P. S. Hall, H. Ghafouri-Shiraz, and D. Wake, Packaging technique for gain enhancement of electrically small antenna designed on gallium arsenide, *Electron. Lett.*, vol. 36, no. 18, pp. 1524–1525, Aug. 2000.

144. G. Bianconi, F. Costa, S. Genovesi, and A. Monorchio, Optimal design of dipole antennas backed by a finite high-impedance screen, *Prog. Electromagn. Res. C*, vol. 18, pp. 137–151, 2011.
145. D. F. Sievenpiper, Forward and backward leaky wave radiation with large effective aperture from an electronically tunable textured surface, *IEEE Trans. Antennas Propag.*, vol. 53, no. 1, pp. 236–247, Jan. 2005.
146. P. Panayi, M. Al-Nuaimi, and I. Ivrissimtzis, Tuning techniques for planar inverted-F antenna, *IEE National Conference on Antennas and Propagation*, York, UK, Mar. 31–Apr. 1, 1999, pp. 259–262.
147. A. Kolsrud, M.-Y. Li, and K. Chang, Dual-frequency electronically tunable CPW-fed CPS dipole antenna, *Electron. Lett.*, vol. 34, no. 7, pp. 609–611, 1998.
148. N. Behdad and K. Sarabandi, Dual-band reconfigurable antenna with a very wide tunability range, *IEEE Trans. Antennas Propag.*, vol. 54, no. 2, pp. 409–416, Feb. 2006.
149. K. Kitatani and S. Yamamoto, Coaxial feed-type microstrip patch antenna with variable antenna height, *Electron. Commun. Jpn. (Part I: Communications)*, vol. 87, no. 2, pp. 10–16, Feb. 2004.
150. J. T. Aberle, S.-H. Oh, D. T. Auckland, and S. D. Rogers, Reconfigurable antennas for wireless devices, *IEEE Antennas Propag. Mag.*, vol. 45, no. 6, pp. 148–154, Dec. 2003.
151. N. Karmakar, Shorting strap tunable stacked patch PIFA, *IEEE Trans. Antennas Propag.*, vol. 52, no. 11, pp. 2877–2884, 2004.
152. N. Behdad and K. Sarabandi, A varactor-tuned dual-band slot antenna, *IEEE Trans. Antennas Propag.*, vol. 54, no. 2, pp. 401–408, Feb. 2006.
153. D. E. Anagnostou, G. Zheng, M. Chryssomallis, J. Lyke, G. E. Ponchak, J. Papapolymerou, and C. G. Christodoulou, Design, fabrication, and measurements of an RF-MEMS-based self-similar reconfigurable antenna, *IEEE Trans. Antennas Propag.*, vol. 54, no. 2, pp. 422–432, Feb. 2006.
154. B. A. Cetiner, H. Jafarkhani, J.-Y. Qian, H. J. Yoo, A. Grau, and F. De Flaviis, Multifunctional reconfigurable MEMS integrated antennas for adaptive MIMO systems, *IEEE Commun. Mag.*, vol. 42, no. 12, pp. 62–70, Dec. 2004.
155. R. Al-Dahleh, C. Shafai, and L. Shafai, Frequency-agile microstrip patch antenna using a reconfigurable MEMS ground plane, *Microw. Opt. Technol. Lett.*, vol. 43, no. 1, pp. 64–67, Oct. 2004.
156. B. A. Cetiner, J. Y. Qian, H. P. Chang, M. Bachman, G. P. Li, and F. De Flaviis, Monolithic integration of RF MEMS switches with a diversity antenna on PCB substrate, *IEEE Trans. Microw. Theory Tech.*, vol. 51, no. 1, pp. 332–335, Jan. 2003.
157. C. W. Jung, M. J. Lee, G. P. Li, and F. De Flaviis, Reconfigurable scan-beam single-arm spiral antenna integrated with RF-MEMS switches, *IEEE Trans. Antennas Propag.*, vol. 54, no. 2, pp. 455–463, Feb. 2006.
158. F. Costa, A. Monorchio, S. Talarico, F. M. Valeri, An active high-impedance surface for low-profile tunable and steerable antennas, *IEEE Antennas Wirel. Propag. Lett.*, vol. 7, pp. 676–680, Sep. 2008.
159. F. Costa, A. Monarchio, and G. P. Vastante, Tunable high-impedance surface with a reduced number of varactors, *IEEE Antennas Wirel. Propag. Lett.*, vol. 10, pp. 11–13, 2011.
160. R. Garg, P. Bhartia, I. Bahl, and A. Ittipiboon, *Microstrip Antenna Design Handbook*. Norwood, MA: Artech House, Norwood, 2001.
161. F. Costa, A. Monorchio, and G. Manara, An equivalent-circuit modeling of high impedance surfaces employing arbitrarily shaped FSS, in *International Conference on Electromagnetics in Advanced Applications (ICEAA '09)*, Turin, Sep. 14–18, 2009, pp. 852–855.
162. F. Costa, S. Genovesi, and A. Monorchio, On the bandwidth of high-impedance frequency selective surfaces, *IEEE Antennas Wirel. Propag. Lett.*, vol. 8, pp. 1341–1344, 2009.

Index